Springer Theses

Recognizing Outstanding Ph.D. Research

For further volumes:
http://www.springer.com/series/8790

Aims and Scope

The series "Springer Theses" brings together a selection of the very best Ph.D. theses from around the world and across the physical sciences. Nominated and endorsed by two recognized specialists, each published volume has been selected for its scientific excellence and the high impact of its contents for the pertinent field of research. For greater accessibility to non-specialists, the published versions include an extended introduction, as well as a foreword by the student's supervisor explaining the special relevance of the work for the field. As a whole, the series will provide a valuable resource both for newcomers to the research fields described, and for other scientists seeking detailed background information on special questions. Finally, it provides an accredited documentation of the valuable contributions made by today's younger generation of scientists.

Theses are accepted into the series by invited nomination only and must fulfill all of the following criteria

- They must be written in good English.
- The topic should fall within the confines of Chemistry, Physics, Earth Sciences, Engineering and related interdisciplinary fields such as Materials, Nanoscience, Chemical Engineering, Complex Systems and Biophysics.
- The work reported in the thesis must represent a significant scientific advance.
- If the thesis includes previously published material, permission to reproduce this must be gained from the respective copyright holder.
- They must have been examined and passed during the 12 months prior to nomination.
- Each thesis should include a foreword by the supervisor outlining the significance of its content.
- The theses should have a clearly defined structure including an introduction accessible to scientists not expert in that particular field.

Christian Flytkjær Jensen

Online Location of Faults on AC Cables in Underground Transmission Systems

Doctoral Thesis accepted by
Aalborg University, Aalborg, Denmark

Springer

Author
Christian Flytkjær Jensen
Department of Energy Technology
Aalborg University
Aalborg
Denmark

Supervisors
Prof. Claus Leth Bak
Department of Energy Technology
Aalborg University
Aalborg
Denmark

Unnur Stella Gudmundsdottir
Transmission Lines
Energinet.dk
Fredericia
Denmark

ISSN 2190-5053 ISSN 2190-5061 (electronic)
ISBN 978-3-319-05397-4 ISBN 978-3-319-05398-1 (eBook)
DOI 10.1007/978-3-319-05398-1
Springer Cham Heidelberg New York Dordrecht London

Library of Congress Control Number: 2014933568

Printed on acid-free paper

Springer is part of Springer Science+Business Media (www.springer.com)

Parts of this thesis have been published in the following articles:

- C. F. Jensen C. L. Bak, U. S. Gudmundsdottir, "State of the art Analysis of Online Fault Location on AC Cables in Underground Transmission Systems", NORD-IS 2011.
- C. F. Jensen, U. S. Gudmundsdottir, C. L. Bak, and A. Abur, "Field Test and Theoretical Analysis of Electromagnetic Pulse Propagation Velocity on Crossbonded Cable Systems", IEEE transaction on power delivery.
- C. F. Jensen, O. M. K. K. Nanayakkara, A. D. Rajapakse, U. S. Gudmundsdottir, and C. L. Bak, "Online Fault Location on Crossbonded Cables Using Sheath Current Signals", IPST 2013, Vancouver, Canada.
- C. F. Jensen, O. M. K. K. Nanayakkara, A. D. Rajapakse, U. S. Gudmundsdottir, and C. L. Bak, "Online Fault Location on Crossbonded Cables Using Sheath Current Signals", Electric Power Systems Research (EPSR) Special IPST 2013 edition.
- C. F. Jensen, U. S. Gudmundsdottir and C. L. Bak, "Online Fault Location on Crossbonded AC Cables in Underground Transmission Systems", Cigré 2014 Paris session.
- C. F. Jensen and C. L. Bak, "Distance Protection of Crossbonded Transmission Cable- Systems", DPSP 2014.

To Nicoline Louisa Frank Iversen

Supervisor's Foreword

I have had the pleasure of following Christian during his studies from the early beginnings at the first semester and from time to time, both as a lecturer and also as a supervisor. Gradually, it became clear to me that he was more talented than the average student, not only in the ability to learn, but also in having a better fundamental understanding of Physics, a profound interest in Electric Power Engineering, the ability to work independently and come up with clever ideas and at the same time being a very nice guy.

Christian had many outstanding presentations during his postgraduate studies. He independently selected a very interesting and complicated ninth semester project related to switching transients in offshore transmission cable connection to a large offshore wind farm. This study was conducted in cooperation with Danish TSO Energinet.dk. His results were of such high quality that they were published in the highly esteemed IPST conference in a scientific paper, 'Switching studies for the Horns Rev 2 wind farm main cable'. A 'normal' student would have selected some kind of similar continuation for the 10th semester master's thesis project in order to continue the success and avoid risk in the final project, but not Christian. He wanted a new technical challenge and so it became. I selected, together with a clever Ph.D. student I supervised at that time, a very challenging project related to the harmonics emitted from large offshore wind farms. This work included many disciplines such as advanced stochastic methods and power system modelling as well as the interpretation and processing of huge amounts of data from real-life measurements. He succeeded again and received the highest possible mark in his final examination. Again, his work was of such high quality that it was published in the Wind Integration Workshop 2011 in a paper entitled 'Probabilistic aspects of harmonic emission of large offshore wind farms'. Christian was a brilliant student and a pleasure to work with, not only seen from a professional point of view, but also from having him as a kind of colleague. I actually faced his immediate parting upon master's with regret. It was a shame to let go of so much clever thinking just in the perfectly right 'engineering mode'.

So what to do as a university man? The only right thing to do is to try offering such a guy a Ph.D. position. In this way, Christian would be able to develop his skills, let them grow and mature, and as a supervisor, I would benefit from accessing and being a part of the highly interesting research work he will perform for 3 years.

Danish TSO Energinet.dk and I have had a very long ongoing and fruitful cooperation. Once, I worked in the practical life transmission engineering business in a regional transmission company where we cooperated much with Energinet.dk (at that time ELSAM), and this continued when I got the position at the Department of Energy Technology, Aalborg University. Many student masters' theses and personal research were initiated by this valuable cooperation. Together, Energinet.dk and I established a research programme 'DANPAC' (DANish Power system with Ac Cables) researching how to use long underground AC cables in the 150 and 400 kV transmission system. One position came up in this programme related to fault location in underground cable systems. I knew Christian would like to continue his research-oriented way of working so I decided to offer him the Ph.D. position. Today, I am happy to say that it turned out to be a very good decision!

In order to understand the motivation of the project and its usefulness, some background information is necessary. Usually, a transmission system consists of overhead lines and only a very limited (and short length) amount of underground cables. The Danish government has decided that almost the entire transmission system has to be undergrounded due to aesthetic reasons. When a fault happens in an overhead line, you can easily find the faulted location and repair it, simply because it is visible. This is not the case with underground cables as they are literally buried and thereby, faults are much more difficult to locate as the cable has to come up from underground for inspection. Therefore, fault location for cable systems is more difficult, time-consuming and expensive compared to overhead lines.

Christian's task was to develop and implement a method capable of finding a fault in an undergrounded cable and with the best possible precision, taking into account the practical limitations a real power system would pose on such a method. In other words, we want to be able, more or less, to dig directly to the fault instead of having to dig up several kilometres of cable.

Christian has solved this very complex task in a very fine and structured way; he worked like a real scientist. However, even more importantly, he always kept full connection with the real world with numerous discussions with me and Energinet.dk, in the end ensuring that his method is actually almost directly applicable to the power system, and Energinet.dk intends to do this for future works. The work was solved using:

- Impedance-based methods for detecting the faulted location
- Travelling wave-based methods for detecting the faulted location
- High quality real-life measurements on the Anholt offshore wind farm 220 kV cable
- Implementation and validation of travelling wave method into Labview environment

Christian's Ph.D. thesis was assessed by highly esteemed Profs. Frede Blaabjerg, Carlo Alberto Nucci and Akihiro Ametani with the designation, 'he is an excellent researcher', a conclusion I strongly support.

When thinking back over the years with Christian, I have come to the conviction that what I liked most when working with him was the profound physics discussions. He is perhaps the student whom I have spent most time supervising, but numerous were our fruitful discussions related to electromagnetic field theory and wave propagations. He could send emails in the evening putting up weird questions like 'Why do we see this bump at the curve here and not when we change?'. The next day, we would use an hour in my office and always come up with the explanation, knowing that we got more and more clever each time we did it. I have very much enjoyed working with Christian during this 3-year period! The book in front of you presents Christian's fine work during his Ph.D. and I sincerely hope you will find it valuable and enjoy reading it.

Aalborg, December 2013 Claus Leth Bak

Preface

This thesis is submitted to the Faculty of Engineering, Science and Medicine at Aalborg University in partial fulfilment of the requirements for the Ph.D. degree in Electrical Engineering. The research has been carried out between 1.09.2010 and 15.07.2013 at the Department of Energy Technology for Energinet.dk by which I was hired as a Ph.D. student for the entire project period.

The project has been followed full time by two supervisors: Prof. Claus Leth Bak (Department of Energy Technology) and Unnur Stella Gudmundsdottir (Energinet.dk).

Energinet.dk has fully funded the research leading to this thesis 'Online Location of Faults on AC Cables in Underground Transmission Systems'. This funding has been vital for this research project. Travelling to conferences, two visits at forging research institutions, renting of laboratory equipment and performance and field measurements were made possible thanks to support from the company.

I spent the period from April to June 2013 at Northeastern University in Boston, USA, under the supervision of Prof. Ali Abur. Here, I worked on two IEEE transaction papers with one of them co-authored by Prof. Abur.

In September 2013, I spent 1 month at the Manitoba HVDC Research Centre in Winnipeg working on fault location on hybrid lines and analysing my field measurements. During this stay, I corroborated with Ph.D. student K. Nanayakkara and Prof. A. D. Rajapakse both from the Department of Electrical and Computer Engineering, University of Manitoba, Winnipeg, Canada.

During the project period, I supervised two master's projects and taught one semester course in power system transients.

The scientific papers written as a part of this Ph.D. are included at the back of both the printed version of the thesis as well as in the PDF file. The papers should not be considered a part of this monograph, but are enclosed if the reader is interested.

This thesis has four parts and appendices. Literature references are presented at the end of every chapter. A list of the authored publications is presented at the end of the thesis. Literature references are shown as [i], where i is the number of the literature in the reference list. Tables, figures and equations are shown as C.F, where C is the chapter number and F is a unique number for the figure or table.

Aalborg, July 2013 Christian Flytkjær Jensen

Acknowledgments

I owe gratitude to many people who helped me in different ways. I would like to thank:

- My supervisors, Claus Leth Bak and Unnur Stella Gudmundsdottir for all their help, comments and many technical discussions during the last 3 years.
- Carl-Erik Madsen, Henrik Kastberg and Carl Willi Hansen for their help with carrying out the field measurements. In the same regards, very special thanks is extended to Filipe Faria da Silva and Unnur Stella Gudmundsdottir for participating during all measurements.
- Filipe Faria da Silva for numerous technical discussions, for reading large parts of the thesis and for many good times in the office.
- Rasmus Schmidt Olsen, Unnur Stella Gudmundsdottir, Joachim Holbøll, Per Balle Holst, Poul Erik Pedersen, Thomas Kvarts, Carsten Rasmussen, Claus Leth Bak and Bjarne Søndergaard Bukh for their contributions and comments during the project period.
- Everyone at the transmission department at Energinet.dk for their contributions.
- Prof. Ali Abur for his hospitality and comments to my work during my 3-month stay at Northeastern University in Boston in 2012.
- Everybody at the Manitoba HVDC Research Centre, Winnipeg, Manitoba for their help with my research during the month I spent there in 2013. Special thanks to John Nordstrom and Juan Carlos Garcia for helping to arrange my stay at the Research Centre at very short notice and for making my stay there pleasant.
- A special thank you to Dr. Jeewantha Da Silva for numerous discussions, valuable advice and guidance throughout the entire project period.
- To Prof. Athula Rajapakse and Ph.D. student Kasun Nanayakkara both from University of Manitoba, Winnipeg, Manitoba for valuable discussions during my stay in Winnipeg.
- To Tine Lykke Tindal Sørensen for proofreading most parts of the thesis.
- Finally, to my girlfriend Nicoline Louisa Frank Iversen for her patience, understanding, love and support especially during the finalising of the thesis.

Contents

Part IV Conclusions

Part I
Preliminaries

Chapter 1
Introduction

A transmission grid is normally laid out as an almost purely overhead line (OHL) network. The introduction of transmission voltage level XLPE cables and the increasing interest in the environmental impact of OHL has resulted in an increasing interest in the use of underground cables on transmission level. In Denmark, the entire 150, 132, and 220 kV as well as parts of the 400 kV transmission network will be placed underground before 2030. The plan of the future Danish transmission system is shown in Fig. 1.1.

Most faults on overhead lines are caused by temporary occurrences as for instance lightning, conductor swing, trees, ice and more. Because the electrical insulation (air) is self-restoring, an auto re-closure method can be used so the system can obtain its original configuration without examining the fault location. When a fault occurs in underground cables, auto re-closure is not used. This is because the insulation material is non-self-restoring, and re-energising the cable without any further actions can very likely lead to more damage on the cable. Instead, the fault must be located and inspected before any action can be taken.

It is in the interest of the system operator to configure the system in such a way that the total system active losses are kept to a minimum. Long outage time of main transmission lines can result in additional losses and bottlenecks because of the non-optimal configurations of the network. Furthermore, production units or consumers connected to a single radial line are disconnected completely from the main grid in case of a fault—this is for instance the case with offshore wind farms.

Off-line fault location time-domain reflectometer—(TDR) and bridge methods can be used directly to locate bolted faults in cable systems [1]. However, it is commonly seen for power cables with extruded insulation that the insulation closes after fault occurrence [1]. The result is a high ohmic fault which can be very difficult to locate using both TDR and bridge methods. Methods which rely on re-opening the insulation at the fault location are therefore used [1]. These can, however, cause more damage to the cable or more seriously fail completely if the equipment used is not powerful enough the re-open the insulation.

C. F. Jensen, *Online Location of Faults on AC Cables in Underground Transmission Systems*, Springer Theses, DOI: 10.1007/978-3-319-05398-1_1,

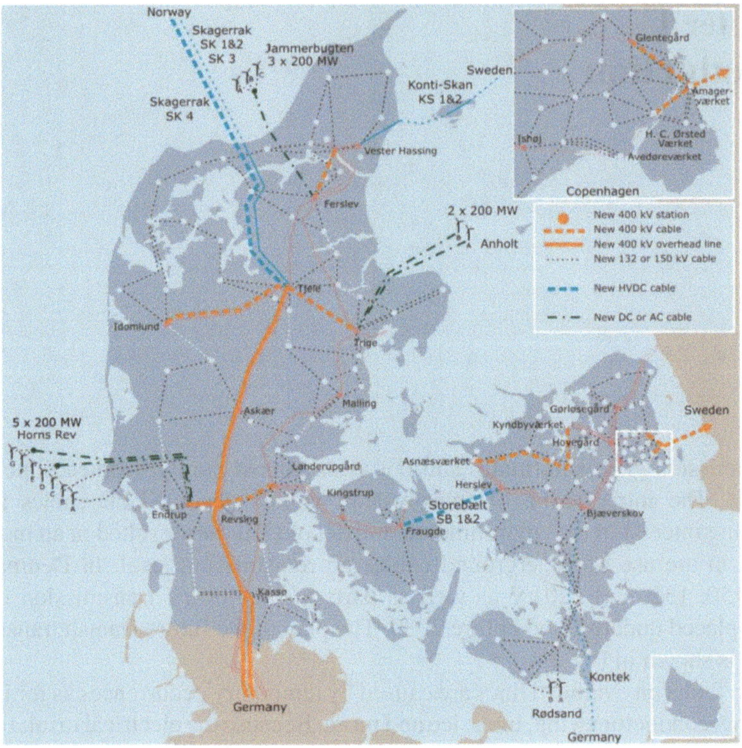

Fig. 1.1 Grid structure planned for Denmark in 2030

On 18 Dec 2002, a single phase to ground fault was detected on the 55 km 150 kV crossbonded cable between the Danish stations Karlsgårde and Blåvand. The cable constitutes the land part of a connection to the offshore wind farm Horns Reef 2.

A local fault location crew were called in and using an off-line surge pulse method, it was attempted to locate the fault. The XLPE insulation had closed after the fault occurrence and the 30 kV surge impulse equipment could not re-open the insulation. More powerful equipment was sent upon request, and a fault location 450 m from Karlsgårde was identified on 19 Dec. An acoustic method was applied and a weak signal could be heard at the predicted location. Joint 30 was after some consideration identified as the faulted joint and on 22 Dec, the replacement of the joint had finished. Unfortunately, a new measurement showed that the cable was still faulted.

On 25 Dec, a high voltage bridge method was used and the fault location was estimated 23.2–23.3 km from Karlsgårde. The day after, a voltage gradient method was used to verify the location whereafter Joint 10 was dug free and replaced. Noon on Sunday, 29 Dec, the repair had finished and the cable was put back into operation. A picture of the faulted cable is shown in Fig. 1.2.

Fig. 1.2 Fault on a single-core cable used for the Horns Reef 2 connection

It took six days to locate the fault and nine days before the line was back in operation. During that period, Energinet.dk, as the Danish transmission system operator, had to compensate the owners of the wind farm. Because the off-line methods are used with difficulty on long crossbonded cables, an online fault location method is desirable.

Online fault location on crossbonded cable systems is in general not studied in detail. Many methods exist and are still being developed for overhead line and cable based distribution systems, but the crossbonding of the sheath at transmission level makes the methods hard to use directly. Furthermore, no high frequency recordings of real-life fault signals on crossbonded cables is available for analysis and verification purposes. Because of this, most research is done on the basis on simulations.

A 400 kV backbone transmission line will connect the biggest Danish substations. This line is planned as mainly an OHL with several short crossbonded cable sections. This backbone line is very important for economical operation of the Danish grid wherefore fault location becomes of importance as well. Because fault location of either crossbonded cables or hybrid lines with crossbonded cables is not studied in detail, Energinet.dk, as the Danish transmission system operator, has decided to sponsor this PhD-project which primary goal is to develop a reliable and accurate method for online faults location on crossbonded AC cables and hybrid lines in transmission systems.

Reference

1. IEEE guide for fault locating techniques on shielded power cable systems. IEEE Std 1234–2007, pp. 1–37 (2007)

Chapter 2
Fault in Transmission Cables and Current Fault Location Methods

The problem formulation of this thesis will depend on already existing fault location methods for crossbonded cables. Therefore, a literature study is conducted and the most important references are presented in the following chapter. Firstly, however, the mechanisms leading to faults in high voltage cables are briefly covered in order to examine which fault location methods are applicable.

2.1 Faults in Transmission Cables

Solid dielectrics, typically cross-linked polyethylene (XLPE) is often used as the main insulation material in high voltage AC-cables today [1]. Internal failures in these cables result from gradual deterioration of the insulation materials between core and sheath [2]. Voids and impurities in the insulation material or between boundaries of different material can initiate a process called treeing leading to insulation breakdown [3].

Electrical trees are formed by locally increased electrical stress and propagate relatively fast in the insulation material until it breaks down. Water trees are another cause of insulation breakdown. They are formed by a local defect and in the presence of moisture, water trees can propagate in the dry insulation under low electrical stress. Water trees have propagated very slowly over the years and are hard to detect as no partial discharges will appear.

When the insulation breaks down, an electric arc forms a low impedance path between the cable's core and sheath. The arc typically burns until the protection system disconnects the cable after the fault is initiated.

At the moment of fault, all internal faults on shielded cables are shunt faults [4]. A low impedance path exists between core and sheath and large fault current flows. When the protection system disconnects the cable, the fault can develop into a series fault or stay as a shunt fault [4]. A combination of both is possible as well. A shunt exists if mechanical forces have ensured a connection between core and sheath, if a

C. F. Jensen, *Online Location of Faults on AC Cables in Underground Transmission Systems*, Springer Theses, DOI: 10.1007/978-3-319-05398-1_2,

carbon-metal bridge exists or if evaporated insulation permits a low resistance path. A series fault is defined as a fault where the conductor is disconnected at one location [4]. This can occur if a part of the conductor or a joint is blown apart at the instance of fault. In case of a shunt fault, two things can happen. The fault can either stay bolted with a solid connection between core and sheath or, as in most case, turn into a fault with a voltage dependent fault resistance [4]. At a low voltage less than 500 V the cable seems non-faulted when measurements are performed from the cable ends. If a voltage larger than 500 V, is applied, flash over at the fault location re-initiates the fault and a fault current can flow.

Internal faults on cables are typically single core to sheath faults. The ground can be included as return path directly from the fault location if the other jacket is damaged by the fault. Two or three phase faults are most often caused by external factors or initiated by a single phase to sheath fault in another cable. The sheath is always involved in any fault type as it encloses the core completely.

Faults in joints will at the moment of fault be shunt faults due to the contact between core and sheath. The core can either have connection to either the sheath of its own cable or to both its own sheath and the transposed sheath. Which sheaths are involved will depend on the type of fault and how is develops. The will affect the different fault location methods differently depending on the way the fault signals are analysed.

2.2 Current Fault Location Methods

In order to identify the most suited fault location methods for crossbonded cables, a review of existing fault location methods is conducted. The current fault location methods for cables can be divided into offline and online methods. The offline methods require special equipment, trained personnel and that the faulted cable is out of service before the methods can be used. The online methods utilise information in the current and voltage measured at the fault locator terminal (FLT) between fault incipience and fault clearance.

The online methods are the main focus in this thesis, but as a general background study, it is of interest to examine the existing offline methods and identify their advantages and weaknesses.

2.2.1 Offline Methods

The current offline methods are thoroughly described in Ref. [4]. The offline methods can be divided into two categories—terminal methods and tracer methods. The terminal methods do, as the name indicates, rely on analysing measurements performed from one or both ends of the cable. The tracer methods rely on the other hand on measurements performed by a trained person walking the cable route. These methods are in general very accurate, but also very manpower- and time consuming. Some of the most common are bridge methods like the Murray-loop, acoustic methods, The Earth Gradient Method and the Magnetic Pickup Method [4]. The tracker methods are used when the online- or offline terminal methods fail.

Several fault location terminal methods are available. The usability of the methods depends on the value of the fault resistance at the fault location.

Most of the terminal methods require a low fault resistance in order to work. If the fault resistance is 5 Ω or below, both TDR and bridge methods can be used directly. The bridge method does not detect the fault and no waves are reflected at the fault location when using TDR methods. To solve the problem, a Surge Arc Reflection Method, Surge pulse reflection method or Burn arc reflection method must be used. These methods rely on temporarily converting the high resistance fault into a low resistance fault. However, IEEE recommends that "Fault-locating techniques that enable fault locating at the lowest possible voltage in the shortest amount of time should be selected" wherefore many of the offline methods are problematic to use [4].

2.2.2 *Online Methods*

The online fault location methods can be subdivided into two primary categories; Impedance- and travelling wave-based methods. As a subcategory of both, knowledge -based methods developed based on fuzzy logic, neural networks and expert systems are proposed. Some optical methods are presented in the literature as well.

Most fault location methods are developed for overhead line transmission systems and distribution systems. Very few publications exist, directly related to fault location on crossbonded cables [5–8]. In the following, the basic concepts for the most commonly used online fault location methods are described.

Impedance-Based Methods

The impedance-based fault location methods compares most often pre-known line parameters to the impedance measured in the case of fault. Based on this comparison the fault location can be estimated.

The line parameters can either be calculated or measured on the transmission line after installation. Often, a representation based on symmetrical components is selected because it can be difficult and time consuming to obtain all components in the series impedance matrix of the line.

Some of the more early single ended methods only utilise the imaginary part of the fault loop impedance for fault location estimation. This is done to omit the influence of the real fault resistance [9, 10]. However, for double sided infeed, the current from the far end source will contribute to the reactance measured by the fault locator (reactance effect) [11]. The impact of the fault resistance on single-terminal fault location methods is a key factor when evaluating their performance.

An early attempt to compensate for the influence of the fault resistance is proposed by Takagi et al. [12]. The line is decomposed into a pre-fault, a pure fault and a superimposed network using the Thevenin theorem. The method assumes the same angle of all line impedances that the line is transposed; that the line parameters are known and that the charging current can be neglected. This assumption is not valid for cables where the charging current can be 20–50 times higher compared to OHLs.

In more recent work, the capacitive effect of the cable is taken into account in for instance [13–15]. The latter two references depend on a commonly used assumption in fault location research; the modal decomposition can be calculated using a real modal transformation matrix (Clarke transformation, e.g.). The proposed real transformation matrix is only valid for fully transposed lines, and errors are introduced if the true frequency dependent modal transformation matrixes are not used as the authors state [15].

The influence of the pre-fault load current is taken into consideration using an iterative process in Refs. [16, 17]. High load currents can be a problem for the single-ended fault location algorithms and must be taken into consideration [18, 19].

Phase coordinate based fault location methods are proposed by authors in Refs. [20, 21]. The methods take into account the unsymmetrical nature of some transmission lines, but require that all self- and mutual impedances are known exactly.

With the development of cheap communication between substations, the two-terminal fault location methods become more widely used [22–26]. Because more information is available for calculating the fault location, the performance of these methods is generally better than single-terminal methods.

The effect of the arc resistance can be eliminated. Often, distributed line models are used where these are based either on symmetrical components or are solved directly in the phase domain. In for instance [24], a two-terminal synchronised method that takes into account line asymmetry, shunt capacitance and fault resistance is setup. This method does, however, assume that all self- and mutual impedances and admittances are known exactly. The method performs well, but the authors point out that additional errors are most likely introduced by the transducers, hardware and errors in the assumed cable parameters.

Several publications discuss the problems associated with the use of current measurements for fault location due the current transformer (CT) saturation [22, 25]. CT saturation can introduce errors when the fundamental phasors are determined from the transient signals recorded at the fault locator terminals. These errors will reflect onto the calculated fault loop impedance and hence the estimation of the fault location.

Some parameter-free fault location methods are described in Refs. [27–29]. These methods rely on estimating the parameters using pre-fault voltages and currents. The methods estimate the line parameters well, but the verification is made with other calculated line parameters. The line parameters do not represent line asymmetry, but an average impedance and admittance is determined.

In Refs. [30, 31] and recently in [32], is it shown that the fault loop impedance of a crossbonded cable is not linear dependent on the fault location. This is due to discrete changes in the zero-sequence impedance at the crossbondings. Errors are introduced for fault location purposes if the commonly used linear assumption between fault location and fault loop impedance is assumed. The references mentioned are based on a protection approach and the effect on fault location is not studied.

Min et al. presents in 2006 and 2007 an impedance-based method which takes into account the crossbonding of the sheath directly [5, 6]. Series impedance matrixes are formulated for each minor section and the fault location is calculated using a distributed representation of the line. The method is tested on a 154 kV 4.491 km underground cable system with five major sections. A maximum error of 0.2038 % is found under the assumption that the faulted major sections as well as all line parameters for each minor section are known. The method is interesting and should be examined further if its assumptions can be proved valid.

Discussion on Impedance-Based Fault Location Algorithms

Several assumptions are made for most impedance-based fault location algorithms. The most common are:

1. The fault loop impedance is linear dependent on the fault location.
2. A sequence representation of the line can be used with no errors or the full series impedance matrix is available and represents the entire line.
3. The fundamental voltage and current phasors can be determined at either one or both cable ends.
4. The influence of the fault resistance, system loading and short circuit power can be eliminated.

The general behaviour of the fault loop impedance and influencing parameters on crossbonded cable systems are not well studied in the literature. References [30–32] give some discussions seen from a protection point of view, but fault location is not considered. In order to evaluate whether an impedance-based fault location method for crossbonded cables is feasible, what accuracy can be obtained and what limitations should be expected, more studies are needed. These are performed later in this thesis.

Travelling Wave Methods

When a fault occurs on a cable system, transient voltage and current waves will travel from the fault location in both directions towards the terminals to where the cable is connected [33]. The basic idea of the travelling wave fault location methods is to identify the arrival instance of one or more of these fault waves and estimate the fault location from the information extracted [34].

The most simple online travelling wave-based method is a single-terminal method. The method relies only on detecting the first and second wave from the fault location as the effective surge impedance of the substation is assumed to be different from the one of the line, such that an incoming wave is reflected back towards the fault. It is also assumed that the fault arc is not extinguished at the fault location so the surge impedance is close to zero, and the wave is almost completely reflected back towards the fault locator terminal [35]. If the arrival instance of the first and second waves at the fault location is captured and the wave velocity is known, the fault location can be estimated as [35]:

$$x = \frac{v_n \cdot \tau_d}{2} \tag{2.1}$$

where v_n is the velocity of a wave of mode n, and τ_d is the time difference between the arrival instance of the two first waves from the fault for a mode n wave. If the fault occurs at more than 50 % of the line length away from the fault location, then τ_d in Eq. (2.1) is calculated as $2\tau_l - (\tau_2 - \tau_1)$ where τ_l is the travelling time for a wave of mode n travelling the entire line length. The method does not rely on a working communication link between two terminals and is therefore a robust solution when it can be used.

A second type is the two-terminal online method where time synchronised data acquired from both ends is used to estimate the fault location. The data from these units can be sent to a common data handling point where the fault location can be determined using Eq. (2.2) [35].

$$x = \frac{l - v_n \cdot \tau_d}{2} \tag{2.2}$$

where v_n is the velocity of a wave of mode n, l is the length of the transmission line, and τ_d is the time difference between the arrival time of the waves at the two fault locator terminals (FLT).

The travelling wave methods rely as shown in Eqs. (2.1) and (2.2) only on knowledge about the wave velocity and on the arrival instance of one or two of the fault created waves at the fault locator terminals. The methods are immune to fault resistance, fault inception angle and network parameters. Furthermore, the same basic method is used for any fault type and works for overhead lines and cables [11].

The idea to analyse travelling waves for fault location on transmission lines was first proposed in 1978 [34]. The work carried out involving fault location on the transmission level is however mainly focused on overhead line systems—for instance [36–40]. Actual experience with travelling wave fault location on a 400 and 132 kV OHL system is presented in Ref. [41].

Travelling wave methods are also widely adopted for fault location on distribution systems [42–45] and to locate high impedance faults [46]. In a recent paper [47], it is suggested that the wave velocity can be eliminated from Eqs. (2.1) and (2.2) by combining the two methods.

The Wavelet Transform (WLT) has over the last years gained a lot of attention for solving fault location problems on transmission lines [48–54]. In for instance [55], the Wavelet Transform is used to detect the arrival instance of the fault created travelling wave for an OHL system. Both a method that requires two-terminal-GPS-synchronised data and a method which uses single ended recordings only are proposed. Research studying different methods for singularity detection using Wavelets has been published [52, 56, 57]. The Lipschitz Exponent transform is a popular measure of singularities in transient signals and is used widely to detect the arrival instance of fault waves [52, 57].

Not much research has been published on fault location directly on crossbonded cables using travelling waves. In 2005 and again in 2007, Jung et al. published an article concerning with the issue of how to discriminate the fault generated travelling waves from the noise in a one ended fault location scheme for crossbonded power cables [7, 8]. The one terminal methods is, according to the authors, preferred because of the simple structure and because the errors associated with the GPS-synchronisation are avoided. The authors introduced a Wavelet-based filtering method that separated the reflections created at the crossbondings from the second wave from the fault location. The method is verified on a 6.284 km cable with six major sections, and errors between 0.08 and 1.8 % relative to the total cable length are obtained (5–111 m). The method seems promising, but it is verified only on a short cable.

Discussion on Travelling Wave-Based Fault Location Algorithms

The travelling wave-based fault location methods are interesting for crossbonded cables. The implementation is more expensive compared to traditional power frequency methods due to the requirements for high frequency data acquisition and a highly accurate common time reference at both line ends. However, the method is simple and independent of many of the system parameters which can affect other fault location algorithms negatively.

Fault location on crossbonded cables using travelling wave methods is not well studied, but is in general considered more complicated compared to fault location on overhead lines and non-crossbonded cables because additional reflections are created at each crossbonding [58–61]. How this affects the use of the single- and two-terminal fault location methods must be studied in further detail before a final evaluation of the method can be made.

Application of Artificial Intelligence for Fault Location

Generally, artificial neural networks (ANN) are used for pattern recognition. The ANN is trained through a number of training cases using a suitable system model to recognise certain behaviour. The capability of the ANN of non-linear mapping, parallel processing and learning makes it useful for fault location if the ANN is given the right input and trained in a proper manner. The ANN type of algorithm is especially useful if no explicit solution can be formulated for the system (multi-ended transmission and distribution systems with laterals).

Fuzzy logic is a non-crisp type of logic that determines relations between objects by soft qualifications. This type of logic is useful for treating ambiguous, vague, imprecise, noisy, or missing input information which can be available for fault location algorithms. Fuzzy logic is often combined with ANN in the fault location schemes proposed in the literature.

Different ANN's must be used for different types of faults because of the different behaviour of the faults. This means that each ANN is trained according to the correct type of fault. The typical inputs for an impedance-based fault location algorithm using artificial networks are pre- and post-fault currents and voltages—also system parameters as loading, short circuit power, etc. can be used.

Most of the theories developed for power system protection and fault location are based on deterministic evaluation schemes [62]. This can give problems because of the complex system models, uncertain determined parameters, the large amount of data that must be processed and changing system configurations. For these reasons, several authors have proposed the use of Fuzzy logic, Neural Networks or a combination of the two to help make the correct decisions in various power system protection problems. In Ref. [63], a single ended fault location algorithm for a 400 kV transmission line is proposed based on neural networks. The input to the ANN are the pre- and post-voltages and currents phasors. The output is the fault resistance and distance to fault. The algorithm is compared to traditional fault location algorithms and it is shown that by correctly training the ANN, it can adapt itself to large variations in the fault resistance and source impedance.

In Ref. [64], a method of accurate fault locator for EHV transmission lines based on radial basis function neural networks is discussed. The locator utilises faulted voltage and current waveforms at one end of the line only.

In Refs. [65, 66], some discussions regarding the structure of neural networks for fault location is presented. In Ref. [65], several different structures are implemented and their performance evaluated.

In Ref. [67], the application of neural networks and Clarke's transformation in fault location on distribution power systems is presented. The locator is able to identify and locate all types of faults with good results. In Ref. [68], application of wavelet fuzzy neural network in locating single line to ground fault (SLG) in distribution lines is discussed. The method is based on post fault transients and steady state measurements. Fuzzy logic and ANN are used to locate the fault. In Refs. [69, 70] fault location on hybrid systems is discussed. The fault locators estimate the fault location well for various hybrid lines. Neural networks are also used for distance protection schemes. The implementation of a neural networks for solving a protection problem is presented in Ref. [71]. Also, fault calcification using neural networks is proposed in the literature. Such a method is presented in Ref. [62] where neural networks are combined with the wavelet transform to classify the fault type. The algorithm is able to classify all types of faults.

Discussion on Application of Artificial Intelligence for Fault Location

The application of artificial networks for fault location on OHL has been discussed by many authors—some work is also published on hybrid systems. No authors have, however, dealt with fault location problems on crossbonded cables. Because the artificial network learns by example (supervised learning) it should, however, be possible to develop such a method using the methods already proposed. The problem is the extensive amount of data needed for training and to ensure that the models used to create the training data are good and reliable. None of the algorithms proposed in the literature are verified in real-life and if the models used to train the artificial networks are oversimplified, good results can be obtained when verifying the algorithm against the same model, but the results will not be useable in real life. How well the most advanced simulation models predict real fault behaviour on crossbonded cable systems must be examined before any final recommendation regarding artificial intelligence methods can be made.

Discussion on State of the Art

The state of the art analysis conducted shows that fault location on crossbonded cables is not a field which is studied in detail. Only few publications are available when considering both impedance and travelling wave-based methods. The publications which are available are centred on very short lines where the lines in the Danish grid will be considerable longer. The use of artificial intelligence for fault location is a relatively new area of research and is mainly focused on OHL systems. Furthermore, not much research that studies the special conditions for crossbonded cable system under faulted conditions is published.

References

1. S. Katakai, Design of XLPE cables and soundness confirmation methods to extra high voltage XLPE cables, in *Transmission and Distribution Conference and Exhibition 2002*, Asia Pacific. IEEE/PES, vol 2, pp. 1411–1415 (2002)
2. B. Koch, Tests on XLPE-insulated cable arcing faults and arc-proofing. IEEE Trans. Power Delivery **3**(4), 1289–1295 (1988)
3. J. Densley, Ageing mechanisms and diagnostics for power cables—an overview. IEEE Electr. Insul. Mag. **17**(1), 14–22 (2001)
4. IEEE guide for fault locating techniques on shielded power cable systems. IEEE Std 1234–2007, pp. 1–37 (2007)
5. S.H. Kang, J.K. Park, S.W. Min, S.R. Nam, W.K. Park, Fault location algorithm for underground power cables using analytical method. *CIGRE General Meeting*, B1–206 (2006)
6. S.-H. Kang, S.-W. Min, S.-R. Nam, J.-K. Park, Fault location algorithm for cross-bonded cables using the singularity of the sheath impedance matrix. Electr. Eng. **89**(7), 525–533 (2007)
7. X.H. Wang, C.K. Jung, J.B. Lee, Y.H. Song, A study on the fault location algorithm on underground power cable system. Electr. Power Syst. Res. **1**, 763–769 (2005)
8. C.K. Jung, J.B. Lee, Fault location algorithm on underground power cable systems using noise cancellation technique, in *Transmission and Distribution Conference and Exposition*, 2008. IEEE/PES, pp. 1–7, April 2008
9. A.A. Girgis, C.M. Fallon, D.L. Lubkeman, A fault location technique for rural distribution feeders. IEEE Trans. Ind. Appl. **29**(6), 1170–1175 (1993)

10. S. Santoso, R.C. Dugan, J. Lamoree, A. Sundaram, Distance estimation technique for single line-to-ground faults in a radial distribution system, in *Power Engineering Society Winter Meeting*, IEEE, vol 4, pp. 2551–2555 (2000)
11. M.M Saha, J. Izykowski, E.Rosolowski, Fault location on power networks (Springer, 2010)
12. T. Takagi, Y. Yamakoshi, M. Yamaura, R. Kondow, T. Matsushima, Development of a new type fault locator using the one-terminal voltage and current data. IEEE Power Eng. Rev. **PER-2**(8), 59–60 (1982)
13. R.H. Salim, K.C.O. Salim, A.S. Bretas, Further improvements on impedance-based fault location for power distribution systems. IET Gener. Trans. Distrib. **5**(4), 467–478 (2011)
14. M. Gilany, E.S.T. El Din, M.M. Abdel Aziz, D.K. Ibrahim, An accurate scheme for fault location in combined overhead line with underground power cable, in *Power Engineering Society General Meeting*, IEEE, vol 3, pp. 2521–2527 (2005)
15. E.S.T.E. Din, M. Gilany, M.M. Abdel Aziz, D.K. Ibrahim, An PMU double ended fault location scheme for aged power cables, in *Power Engineering Society General Meeting*, IEEE, vol 1, pp. 80–86 (2005)
16. S.-J. Lee, M.-S. Choi, S.-H. Kang, B.-G. Jin, D.-S. Lee, B.-S. Ahn, N.-S. Yoon, H.-Y. Kim, S.-B. Wee, An intelligent and efficient fault location and diagnosis scheme for radial distribution systems. IEEE Trans. Power Delivery **19**(2), 524–532 (2004)
17. E.C. Senger, G. Manassero, C. Goldemberg, E.L. Pellini, Automated fault location system for primary distribution networks. IEEE Trans. Power Delivery **20**(2), 1332–1340 (2005)
18. L. Eriksson, M.M. Saha, G.D. Rockefeller, An accurate fault locator with compensation for apparent reactance in the fault resistance resulting from remore-end infeed. IEEE Trans. Power Apparatus Syst. **PAS-104**(2), 423–436 (1985)
19. M.T. Sant, Y.G. Paithankar, Online digital fault locator for overhead transmission line. Proc. Inst. Electr. Eng. **126**(11), 1181–1185 (1979)
20. M.M. Saha, J. Izykowski, E. Rosolowski, B. Kasztenny, A new accurate fault locating algorithm for series compensated lines. IEEE Trans. Power Delivery **14**(3), 789–797 (1999)
21. M.M. Saha, K. Wikstrom, J. Izykowski, E. Rosolowski, Fault location in uncompensated and series-compensated parallel lines, in *Power Engineering Society Winter Meeting*, IEEE, vol 4, pp. 2431–2436 (2000)
22. S.M. Brahma, A.A. Girgis, Fault location on a transmission line using synchronized voltage measurements. IEEE Trans. Power Delivery **19**(4), 1619–1622 (2004)
23. V. Cook, Fundamental aspects of fault location algorithms used in distance protection. IEE Proc. C Gener. Transm. Distr. **133**(6), 359–368 (1986)
24. A.T. Johns, S. Jamali, Accurate fault location technique for power transmission lines. IEE Proc. C Gener. Transm. Distr. **137**(6), 395–402 (1990)
25. M.M. Saha, J. Izykowski, E. Rosolowski, A two-end method of fault location immune to saturation of current transformers, in *8th IEE International Conference on Developments in Power System Protection*, vol 1, pp. 172–175 (2004)
26. R. Das, M.S. Sachdev, T.S. Sidhu, A fault locator for radial subtransmission and distribution lines, in *Power Engineering Society Summer Meeting*, IEEE, vol 1, pp. 443–448 (2000)
27. G. Preston, Z.M. Radojević, C.H. Kim, V. Terzija, New settings-free fault location algorithm based on synchronised sampling. IET Gener. Transm. Distr. **5**(3), 376–383 (2011)
28. S. Asgarifar, M.T. Hagh, M.M. Hosseini, A novel fault location algorithm for double fed distribution networks, in *Power Engineering and Automation Conference (PEAM)*, IEEE, vol 1, pp. 327–330 (2011)
29. P. Dawidowski, J. Izykowski, A. Nayir, Non-iterative algorithm of analytical synchronization of two-end measurements for transmission line parameters estimation and fault location, in *7th International Conference on Electrical and Electronics Engineering (ELECO)*, pp. I-76–I-79 (2011)
30. V. Leitloff, X. Bourgeat, G. Duboc, Setting constraints for distance protection on underground lines, in *7th International Conference on (IEE) Developments in Power System Protection*, pp. 467–470 (2001)

31. D.A. Tziouvaras, Protection of high-voltage AC cables, in *Power Systems Conference: Advanced Metering, Protection, Control, Communication, and Distributed Resources, PS '06*, pp. 316–328 (2006)
32. Teruo Ohno, *Dynamic Study on the 400 kV 60 km Kyndbyvarrket Asnarsvarrket Line* (Energinet.dk, 1st edn, 2013)
33. A. Greenwood, *Electrical Transients in Power Systems* (Wiley, 2nd edn, 1991), ISBN 0-471-62058-0
34. H.W. Dommel, J.M. Michels, High speed relaying using traveling wave transient analysis, *IEEE PES Winter Power Meeting* (1978)
35. IEEE guide for determining fault location on ac transmission and distribution lines. IEEE Std C37.114-2004, pp. 1–36 (2005)
36. M. Coury, D.V.F. da Silva, M. Oleskovicz, A fault locator for three-terminal lines based on wavelet transform applied to synchronized current and voltage signals. *Transmission and Distribution Conference and Exposition: Latin America, TDC '06*, IEEE/PES, pp. 1–6 (2006)
37. G.W. Philips, S.M. Chin, M.A. Redfern, F. Jiang, Z.Q. Bo, A GPS based fault location scheme for distribution line using wavelet transform technique. *IPST 99—International Conference On Power System Transients, New Protection Techniques*, pp. 224–228 (1999)
38. M.A. Ramos, J.S. Filho, J.M. Beck, S.L. Zimath, N. Mueller, A GPS based fault location scheme for distribution line using wavelet transform technique. *63rd Annual Conference on Protective Relay Engineers*, pp. 1–7 (2010)
39. M. Han, C. Lee, H. Park, H. Jung, Y. Park, M. Shin, Novel technique for fault location estimation on parallel transmission lines using wavelet. Electr. Power Energy Syst. **29**, 76–82 (2007)
40. W. Threevithayanon, N. Hoonchareon, Fault data synchronization using wavelet for improving two-terminal fault location algorithm. *Power and Energy Engineering Conference (APPEEC)*, Asia-Pacific, pp. 1–4 (2010)
41. M.K. Hoi, Traveling wave fault locator experience on CLP power transmission network. *CEPSI Conference*, Fukuoka, Japan, T2-B-8 (2002)
42. C.A. Nucci, M. Polone, L. Peretto, A. Borghetti, S. Corsi, R. Tinarelli, On the use of continues-wavelet transform for fault location in distribution power systems. Electr. Power Energy Syst. **28**, 608–617 (2006)
43. M. Paolone, A. Borghetti, M. Bosetti, A. Abur, Integrated use of time-frequency wavelet decompositions for fault location in distribution networks: Theory and experimental validation. *International Conference on Power Systems Transients (IPST)*, p. 313 (2009)
44. M. Di Silvestro, C.A. Nucci, A. Borghetti, M. Bosetti, M. Paolone, Continuous-wavelet transform for fault location in distribution power networks: Definition of mother wavelets inferred from fault originated transients. *International Conference on Power Systems Transients (IPST)*, p. 22 (2007)
45. A.S. Bretas, R.H. Salim, K.R. Caino de Oliveira, Fault detection in primary distribution systems using wavelets. *International Conference on Power Systems Transients (IPST)* (2007)
46. E.M. Aboul-Zahab, S.M. Saleh, D.K. Ibrahim, E.S.T. Eldin, Real time evaluation of DWT-based high impedance fault detectioin in EHV transmision. Electr. Power Syst. Res. **80**, 907–914 (2010)
47. M. Gilany, D.K. Ibrahim, E.S.T. Eldin, Traveling-wave-based fault-location scheme for multiend-aged underground cable system. IEEE Trans. Power Delivery **22**(1), 82–89 (2007)
48. M. Paolone, A. Borghetti, C.A. Nucci, An automatic system to locate phase-to-ground faults in medium voltage cable networks based on the wavelet analysis of high-frequency signals, in *PowerTech 2011*, IEEE, Trondheim, pp. 1–7 (2011)
49. A. Borghetti, M. Bosetti, C.A. Nucci, M. Paolone, A. Abur, Integrated use of time-frequency wavelet decompositions for fault location in distribution networks: theory and experimental validation. IEEE Trans. Power Delivery **25**(4), 3139–3146 (2010)
50. A. Borghetti, M. Bosetti, M. Di Silvestro, C.A. Nucci, M. Paolone, Continuous-wavelet transform for fault location in distribution power networks: definition of mother wavelets inferred from fault originated transients. IEEE Trans. Power Systems **23**(2), 380–388 (2008)

51. F.Z. He, S.L. Ling, Z. Bo, Fault detection and classification in EHV transmission line based on wavelet singular entropy. IEEE Trans. Power Delivery **25**(4), 2156–2163 (2010)
52. S. Lin, Z.Y. He, X.P. Li, Q.Q. Qian, Travelling wave time-frequency characteristic-based fault location method for transmission lines. IET Gener. Transm. Distrib. **6**(8), 764–772 (2012)
53. F.B. Costa, B.A. Souza, N.S.D. Brito, Effects of the fault inception angle in fault-induced transients. IET Gener. Trans. Distr. **6**(5), 463–471 (2012)
54. M. Korkali, A. Abur, Fault location in meshed power networks using synchronized measurements. *North American Power Symposium (NAPS)***2010**, 1–6 (2010)
55. F.H. Magnago, A. Abur, Fault location using wavelet. IEEE Trans. Power Delivery, **13**(4), 1475–1480 (1998)
56. S. Mallat, W.-L. Hwang, Singularity detection and processing with wavelets. IEEE Trans. Inf. Theory **38**(2), 617–643 (1992)
57. L. Ke, W. Houjun, A novel wavelet transform modulus maxima based method of measuring lipschitz exponent, in *International Conference on Communications, Circuits and Systems, ICCCAS 2007*, pp. 628–632 (2007)
58. N. Nagaoka, A. Ametani, Transient calculations on crossbonded cables. IEEE Trans. Power Apparatus Syst. **PAS-102**(4), 779–787 (1983)
59. Y. Itoh, N. Nagaoka, A. Ametani, Transient analysis of a crossbonded cable system underneath a bridge. IEEE Trans. Power Delivery **5**(2), 527–532 (1990)
60. U.S. Gudmundsdottir, B. Gustavsen, C.L. Bak, W. Wiechowski, Field test and simulation of a 400-kV cross-bonded cable system. IEEE Trans. Power Delivery **26**(3), 1403–1410 (2011)
61. U.S. Gudmundsdottir, Modeling of long High Voltage AC cables in Transmission Systems. (Energinet.dk, 2nd edn, 2010) ISBN 978-87-90707-73-6
62. O.A.S. Youssef, Combined fuzzy-logic wavelet-based fault classification technique for power system relaying. IEEE Trans. Power Delivery **19**(2), 582–589 (2004)
63. Z. Chen, J.-C. Maun, Artificial neural network approach to single-ended fault locator for transmission lines. IEEE Trans. Power Systems **15**(1), 370–375 (2000)
64. M. Joorabian, S.M.A. Taleghani, S. Asl, R.K. Aggarwal, Accurate fault locator for EHV transmission lines based on radial basis function neural networks. Electr. Power Syst. Res. **71**(3), 195–202 (2004)
65. J. Gracia, A.J. Mazon, I. Zamora, Best ANN structures for fault location in single-and double-circuit transmission lines. IEEE Trans. Power Delivery **20**(4), 2389–2395 (2005)
66. S. Osowski, R. Salat, Fault location in transmission line using hybrid neural network. Int. J. Comput. Math. Electr. Electron. Eng. **21**, 18–30 (2002)
67. L.S. Martins, J.F. Martins, V.F. Pires, C.M. Alegria, The application of neural networks and Clarke–Concordia transformation in fault location on distribution power systems, vol 3, pp. 2091–2095 (October 2002)
68. F. Chunju, K.K. Li, W.L. Chan, Y. Weiyong, Z. Zhaoning, Application of wavelet fuzzy neural network in locating single line to ground fault (SLG) in distribution lines. Int. J. Electr. Power Energy Syst. **29**(6), 497–503 (2007)
69. J. Sadeh, H. Afradi, A new and accurate fault location algorithm for combined transmission lines using adaptive network-based fuzzy inference system. Electr. Power Syst. Res. **79**(11), 1538–1545 (2009)
70. C.K. Jung, K.H. Kim, J.B. Lee, Bernd Klöckl, Wavelet and neuro-fuzzy based fault location for combined transmission systems. Int. J. Electr. Power Energy Syst. **29**(6), 445–454 (2007)
71. P.K. Dash, A.K. Pradhan, G. Panda, A novel fuzzy neural network based distance relaying scheme. IEEE Trans. Power Delivery **15**(3), 902–907 (2000)

Chapter 3
Problem Formulation and Thesis Outline

The main objective of this thesis is to develop one or several methods for accurate fault location in underground crossbonded cable networks and hybrid systems. A thorough investigation into the behaviour of the crossbonded cable system under steady state and transient faulted conditions must be carried out. An important part of the work will be to verify the developed fault location method against field measurements of faults emulated on an installed crossbonded cable.

As the developed fault location method has to be practically applicable, a study of the required measuring equipment must be performed. This includes both the measuring equipment for the field tests, but also specifications for the equipment that must be installed in the substations for the developed fault location method to work.

Fault location using impedance-based methods on crossbonded cable systems is not well studied in the literature. However, the method is of special interest as it is cheap and fault writers for power frequency signals are already installed in all Danish substations. A good accuracy can be obtained on overhead lines and distribution systems if all line parameters are known. It must be examined whether this is the case with crossbonded cables as well.

The travelling wave methods seem promising for fault location on cables in general as they are immune to many of the parameters which makes fault location using impedance-based methods difficult. The use of the travelling wave method is, however, not studied in detail for crossbonded cables either. The crossbondings introduce new difficulties and their effect on the fault created travelling waves must be examined.

The fuzzy logic, neural networks and expert systems all rely on some sort of training and are therefore dependent on a model which is able to predict the behaviour of the crossbonded cable system in faulted conditions with high accuracy.

In recent years, power system simulation programs which allow an easy and precise implementation of models for crossbonded cables have become commercially available for both steady state and transient analysis.

C. F. Jensen, *Online Location of Faults on AC Cables in Underground Transmission Systems*, Springer Theses, DOI: 10.1007/978-3-319-05398-1_3,

Due to the destructive nature of faults on power system equipment, almost all research in the area of fault location is based on simulations. To prove the validity of the current models, it is of interest to compare results obtained using these models to field measurements of various faults on crossbonded cables. This will show whether results for fault location studies based on simulations can be trusted or not. This is one of the main contribution of the work presented in this thesis.

With the developments in consumer and industrial electronics, data acquisition and online computing is made easy and cheap. Computational heavy algorithms can be implemented in real-time environments at a low cost. This allows the use of new fault location algorithms in fault locator units installed at numerous locations in the transmission network. Furthermore, fast and reliable data communication is cheap and widely available.

Based on the discussion above, the main objectives of this thesis are formulated as:

1. To study the steady state and transient behaviour of a crossbonded cable system under fault conditions.
2. To analyse the correct modelling and simulation techniques of fault location studies on crossbonded cable systems.
3. To identify and possibly improve the best suited fault location method for crossbonded cables and hybrid lines.
4. To study the necessary measuring equipment required for sufficient accurate fault location in crossbonded cable networks and on hybrid lines.
5. To verify the proposed fault location method using field measurements conducted on a real-life crossbonded cable.
6. To develop a prototype of the proposed fault locator system.

Energinet.dk plans to install the developed fault locator unit in the Danish substations, so it is important that economical aspects are included in the work. Methods which are too expensive to realise in practise are therefore not of interest.

3.1 Thesis Outline

This thesis is divided into five parts. These are:

I - Preliminaries
The preliminaries contain an introduction to the thesis and a review of both the causes of faults in high voltage cables and the current off- and online fault location techniques. The problem formulation of the thesis is defined and the outline of the thesis is given.

II - Fault location on crossbonded cables using impedance-based methods
This part is divided into two chapters. Chapter 4 presents the series phase- and sequence impedance matrices of crossbonded cable systems. The fault loop

impedance is determined for a single phase to sheath fault and the parameters influencing the impedance measured at the fault locator terminals are identified and their effects studied. Fault location on hybrid lines using impedance-based methods is discussed and the limitations identified.

One of the major contributions of this thesis is providing field measurements for verification of fault location methods for crossbonded cables. In Chap. 5, impedance-based field measurements conducted on the electrical connection to the offshore wind park Anholt are presented. The measurements are used to evaluate the performance of the most current cable models for steady state analysis and thus evaluate whether the impedance-based fault location methods are useable for crossbonded cables. The results from the comparison will also show whether the neural networks can be trained using output from the cable models.

III - Fault Location on Crossbonded Cables using Travelling Waves

This part is divided into seven chapters. First, electromagnetic wave propagation on a three single-core solid-bonded and crossbonded cable is examined in Chap. 6. Chapter 7 studies the use of the single and two-terminal fault location methods on crossbonded cables. In Chap. 8, the parameters influencing a two-terminal fault location method for fault location on crossbonded cables are examined. Chapter 9 examines fault location on different line type using online travelling wave methods.

Travelling wave-based field measurements for verification of fault location methods for crossbonded cables are presented in Chap. 10. These are also conducted on the Anholt electrical connection. In Chap. 11, use of the Wavelet Transform for fault location on crossbonded cable systems examined. A theoretical examination of the transform with specific focus on its limitations. In the last chapter of part III, Chap. 12, the development process of a fault locator system for crossbonded cables is described.

IV - Conclusions

In the last part, the final conclusions for the thesis are given. The contributions are listed and the publications based on the thesis are presented. Furthermore, the future work is described.

V - Appendix

The thesis contains four appendices. One presenting the impedance-based fault location measurement results, one where the power system components used in the thesis are presented, an appendix that demonstrates how positive and zero-sequence impedances can be measured on three-phase transmission lines and one showing the single line diagram of GIS-station Karstrup.

Part II
Fault Location on Crossbonded Cables Using Impedance-Based Methods

Online fault location using impedance-based methods are very popular and widely used for overhead lines. Several single- and two-terminal methods exist where the two-terminal methods can be sub-divided into either un- or synchronised methods. Methods for fault location in cables on the distribution level do exist, but very few publications are available on crossbonded cable systems. Most impedance-based methods rely on a comparison of the measured fault loop impedance to pre-known line parameters. Thus, a correct representation of these parameters is necessary for correct results.

Many fault location methods are based on symmetrical components. These methods assume that the phase domain quantities can be transformed to a decoupled domain where the fault location can be estimated. Some models take into account the distributed nature of transmission lines, but very few take into account the complicated return path of the fault current in the ground and sheath system of a crossbonded cable.

Artificial intelligence impedance-based methods rely on models that provide good training data. In this part, the fault loop impedance of a crossbonded cable system is studied in detail. Calculation methods, physical behaviour and influencing parameters are examined. To evaluate the performance of the most current cable models, field measurements are carried out on the electrical connection to the offshore wind farm Anholt. These studies lead to a recommendation whether the impedance-based fault location methods can be used on crossbonded cables or not.

Part II
Fault Location on Crossbonded Cables Using Impedance-Based Methods

[illegible]

Chapter 4
Series Phase and Sequence Impedance Matrices of Crossbonded Cable Systems

In this section, the series impedance matrix of a crossbonded cable system is discussed with special attention to the issues of interest for fault location. The phase series impedance matrix describing one minor section is comprised of the cable conductors' self and mutual impedances with earth return. For high frequency studies, the calculations of these impedances involve complex calculations using Bessel's functions - some of the most important contributions are [1–5]. At power frequency, simplification can, however, be made [6]. Based on these simplified formulas, the sequence components can be calculated using the classical method proposed by Fortescue [7]. However, deriving formulas for the whole cable system is much more complicated as the sheath bonding, the field and substation grounding resistance need to be included [8]. Before the impedance matrices are discussed, a case study single-core cable which will be used throughout the rest of the report is introduced.

4.1 The Single-Core Case Study Cable

The single-core cable introduced in this section will throughout the rest of this thesis be used as a case study cable. The case study cable is a 165 kV single-core cable produced by ABB and is used to connect the Danish offshore wind park Horns Reef 2 to the main grid. The construction of the cable is shown in Fig. 4.1a.

The cable's core has a cross section of 1200 mm^2 and a composite metal sheath made from wound copper wires and an laminated aluminium sheath. The parameters given by the manufacturer are presented in Table 4.1.

Some simplifications are made when high voltage cables are impplemented using the models available today. Solid conductors of one material and loss-less insulation materials are assumed. Furthermore, very often no semi-conductive layers are included in the model. The model implemented in most simulation programs and the model used in this thesis is presented in Fig. 4.1b.

C. F. Jensen, *Online Location of Faults on AC Cables in Underground Transmission Systems*, Springer Theses, DOI: 10.1007/978-3-319-05398-1_4,

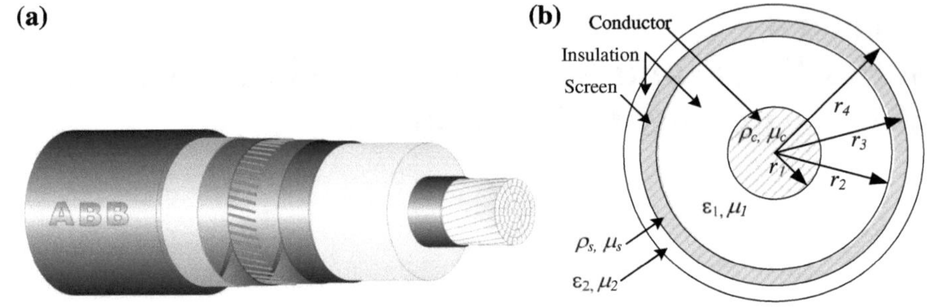

Fig. 4.1 **a** Construction of ABB 165 kV case study cable used in this thesis, and **b** Cable model used in this thesis

Table 4.1 ABB land cable parameters

Parameter	Value
Conductor aluminium [mm^2]	1200
Conductor outer diameter [mm]	41.5
Conductor screen thickness [mm]	1.5
Insulation diameter [mm]	79
Insulation screen thickness [mm]	1.0
Longitudinal water barrier [mm]	0.6
Copper wire screen cross section [mm^2]	95
Longitudinal water barrier thickness [mm]	0.6
Radial water barrier thickness [mm]	0.2
Outer cover diameter [mm]	95

Table 4.2 Model input data for 165 kV single-core case study cable used in this thesis

Parameter	Value
r_1	20.75 mm
r_2	40.5 mm
r_3	41.61 mm
r_4	47.5 mm
ϵ_1	2.74
ϵ_2	2.3
ρ_c	$3.156 \cdot 10^{-8}$ Ω·m
ρ_s	$3.65 \cdot 10^{-8}$ Ω·m

In order to compensate for the lack of complexity in the cable model, the input parameters given to the model are adjusted. This is done for the ABB cable by using the methods presented in [9–11]. The calculations are shown in [12] and the results are presented in Table 4.2.

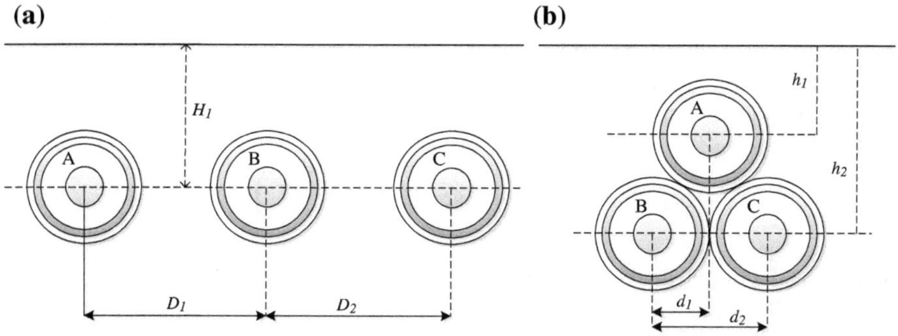

Fig. 4.2 Cable formations used in the Danish transmission grid **a** Flat formation and **b** touching trefoil formation. In Fig. (**a**) *H1* is the depth of the cable system in flat formation, *D1* is the distance between cable A and B, *D2* is the distance between cable B and C. In Fig. (**b**) *h1* is the depth of cable A, *h2* is the depth of cable B and C, *d1* is the distnce betwen cable A and B and *d2* is the distance between cable B and C

Two laying configurations are studied in this thesis. These are the flat formation and the touching trefoil formation. Both formations will be used in the future Danish transmission network and their definitions are shown in Fig. 4.2a and b.

Throughout the rest of this thesis, cables laid in flat formation are buried 1.3 m underground ($H_1 = 1.3$) with a separation of 0.4 m ($D_1 = D_2 = 0.4$) m. For a touching trefoil formation (Fig. 4.2b), h_1 is 1.3 m and $h_2 = \sqrt{(2r_4)^2 - r_4^2} =$ 1.3823 m. The distance d_1 is equal the radius of the cable r_4 and d_2 is equal $2 \cdot r_4$. The ground resistivity is assumed to be homogenous and equal to 100 Ωm.

4.2 Series Impedance Matrix

The longitudinal voltage drop per unit length of a three-phase cable system with single-core coaxial cables and no armour is related to the core and sheath currents as shown in Eq. 4.1.

$$\begin{bmatrix} \underline{U}_{C_A} \\ \underline{U}_{C_B} \\ \underline{U}_{C_C} \\ \hline \underline{U}_{S_A} \\ \underline{U}_{S_B} \\ \underline{U}_{S_C} \end{bmatrix} = \left[\begin{array}{ccc|ccc} \underline{Z}_{CACA} & \underline{Z}_{CACB} & \underline{Z}_{CACC} & \underline{Z}_{CASA} & \underline{Z}_{CASB} & \underline{Z}_{CASC} \\ \underline{Z}_{CBCA} & \underline{Z}_{CBCB} & \underline{Z}_{CBCC} & \underline{Z}_{CBSA} & \underline{Z}_{CBSB} & \underline{Z}_{CBSC} \\ \underline{Z}_{CCCA} & \underline{Z}_{CCCB} & \underline{Z}_{CCCC} & \underline{Z}_{CCSA} & \underline{Z}_{CCSB} & \underline{Z}_{CCSC} \\ \hline \underline{Z}_{SACA} & \underline{Z}_{SACB} & \underline{Z}_{SACC} & \underline{Z}_{SASA} & \underline{Z}_{SASB} & \underline{Z}_{SASC} \\ \underline{Z}_{SBCA} & \underline{Z}_{SBCB} & \underline{Z}_{SBCC} & \underline{Z}_{SBSA} & \underline{Z}_{SBSB} & \underline{Z}_{SBSC} \\ \underline{Z}_{SCCA} & \underline{Z}_{SCCB} & \underline{Z}_{SCCC} & \underline{Z}_{SCSA} & \underline{Z}_{SCSB} & \underline{Z}_{SCSC} \end{array}\right] \begin{bmatrix} \underline{I}_{C_A} \\ \underline{I}_{C_B} \\ \underline{I}_{C_C} \\ \hline \underline{I}_{S_A} \\ \underline{I}_{S_B} \\ \underline{I}_{S_C} \end{bmatrix} \tag{4.1}$$

where $\underline{U}_{C_i}$ and $\underline{I}_{C_i}$ are the voltage and current of core i, $\underline{U}_{S_i}$ and $\underline{I}_{S_i}$ are the voltage and current of sheath i, $\underline{Z}_{CiCi}$ is the self-impedance of the core of cable i, $\underline{Z}_{SiSi}$ is the self-impedance of the sheath of cable i, $\underline{Z}_{CiCm}$ is the mutual impedance between core of cable i and the core of cable m, $\underline{Z}_{SiSm}$ is the mutual impedance between sheath of cable i and the sheat of cable m and $\underline{Z}_{CiSm}$ is the mutual impedance between core of cable i and the sheath of cable m.

For power frequency studies, the self-impedance of a solid conductor with earth return is determined as shown in Eq. 4.2 [13].

$$\underline{Z}_{CC} = Rc_{ac} + \pi^2 10^{-4} f + j4\pi 10^{-4} f \ln\left(\frac{D_e}{r_1}\right) \qquad [\Omega/\text{km}] \tag{4.2}$$

where r_1 is the outer radius of the core, Rc_{ac} is the AC resistance of the core, f is the frequency, D_e is the depth of the equivalent earth return conductor. D_e is calculated as shown in Eq. 4.3 [13].

$$D_e = 658.87\sqrt{\frac{\rho_e}{f}} \qquad [\text{m}] \tag{4.3}$$

where ρ_e is the resistivity of the earth.

The self-impedance of the sheath conductor with earth return is determined as shown in Eq. 4.4 [13].

$$\underline{Z}_{SS} = Rs_{ac} + \pi^2 10^{-4} f + j4\pi 10^{-4} f \left[\frac{\mu_S}{4} \cdot f(r_3, r_2) + \ln\left(\frac{D_e}{r_3}\right)\right] \quad [\Omega/\text{km}] \tag{4.4}$$

where r_2 and r_3 are the inner and outer radius of the sheath, Rs_{ac} is the AC resistance of the sheath and μ_S is the permeability of the sheath material.

The term $f(r_3, r_2)$ in Eq. 4.4 is calculated as [13]:

$$f(r_3, r_2) = 1 - \frac{2r_2^2}{(r_3^2 - r_2^2)} + \frac{4r_2^4}{(r_3^2 - r_2^2)^2} \qquad [-] \tag{4.5}$$

The mutual impedance between core or sheath of cable i and core or sheath of cable j with earth return is calculated as [13]:

$$\underline{Z}_{ij} = \pi^2 10^{-4} f + j4\pi 10^{-4} f \cdot \ln\left(\frac{D_e}{S_{ij}}\right) \qquad [\Omega/\text{km}] \tag{4.6}$$

The series impedance matrix shown in Eq. 4.1 is redefined as shown in Eq. 4.7.

$$\mathbf{Z}_{abc} = \left[\begin{array}{c|c} \mathbf{Z}_C & \mathbf{Z}_{CS} \\ \hline \mathbf{Z}_{CS}^T & \mathbf{Z}_S \end{array}\right] \qquad [\Omega] \tag{4.7}$$

where $\mathbf{Z}_{CS}^T$ is the transpose of $\mathbf{Z}_{CS}$. The symmetrical components' transformation is calculated as shown in Eq. 4.8.

$$\mathbf{Z}_{012} = \mathbf{T}^{-1}\mathbf{Z_{abc}}\mathbf{T} \qquad [\Omega] \tag{4.8}$$

where the $\mathbf{T}$ is defined as:

$$\mathbf{T} = \begin{bmatrix} \mathbf{A} & 0 \\ 0 & \mathbf{A} \end{bmatrix} \tag{4.9}$$

and $\mathbf{A}$ in Eq. 4.9 is defined as shown in Eq. 4.10.

$$\mathbf{A} = \begin{bmatrix} 1 & 1 & 1 \\ 1 & \underline{a}^2 & \underline{a} \\ 1 & \underline{a} & \underline{a}^2 \end{bmatrix} \tag{4.10}$$

with the a operator being a complex number equal $\underline{a} = \exp(j2\pi/3) = -\frac{1}{2} + j\frac{\sqrt{3}}{2}$.

The sub-matrices $\mathbf{Z}_C$, $\mathbf{Z}_S$ and $\mathbf{Z}_{CS}$ defined in Eq. 4.7 all become balanced (equal diagonal and off-diagonal elements) if the cable section is laid in trefoil. This occurs because the distance from centre to centre and thereby the mutual coupling between core/core, sheath/sheath and core and the sheath to any of the two other cables are equal. For the system laid in trefoil, the resulting 6×6 sequence matrix is on the form:

$$\mathbf{Z}_{012,6\times6} = \begin{bmatrix} \underline{Z}_{0,n} & 0 & 0 & \underline{Z}_{0m} & 0 & 0 \\ 0 & \underline{Z}_{1,n} & 0 & 0 & 0 & 0 \\ 0 & 0 & \underline{Z}_{2,n} & 0 & 0 & 0 \\ \underline{Z}_{0m} & 0 & 0 & \underline{Z}_{0,n+1} & 0 & 0 \\ 0 & 0 & 0 & 0 & \underline{Z}_{1,n+1} & 0 \\ 0 & 0 & 0 & 0 & 0 & \underline{Z}_{2,n+1} \end{bmatrix} \quad [\Omega] \tag{4.11}$$

where $\underline{Z}_{0,n}$, $\underline{Z}_{1,n}$ and $\underline{Z}_{2,n}$ are the zero, positive and negative sequence impedance for the nth circuit, and $\underline{Z}_{0m}$ is the zero sequence mutual impedance. The sequence components are decoupled because the off-diagonal are zeros. The positive, negative and zero-sequence components can therefore by used to describe the cable placed in trefoil with no loss of accuracy. This is not the case for a system in flat formation. Here, the use of only the diagonal sequence impedances will lead to errors if the sequence parameters are used directly in fault location algorithms.

For cable sections with solidly bonded and grounded sheaths at each cable end, the sheath voltages can be eliminated and the 6×6 matrix reduced to a 3×3 matrix as shown in Eq. 4.12.

$$\mathbf{Z}_{abc,3\times3} = \mathbf{Z}_C - \mathbf{Z}_{CS}\mathbf{Z}_S^{-1}\mathbf{Z}_{CS}^T \quad [\Omega] \tag{4.12}$$

With eliminated sheaths, the sequence impedance can be calculated shown in Eq. 4.13.

$$\mathbf{Z}_{012,3\times3} = \mathbf{A}^{-1}\mathbf{Z}_{\mathbf{abs}}\mathbf{A} \quad [\Omega] \tag{4.13}$$

For trefoil formations, the off-diagonal elements of $\mathbf{Z}_{012,3\times3}$ are zero where, if a flat formation is used, this is generally not the case. The errors are, however, generally small and can in most cases be neglected [14].

4.2.1 Impedance Matrix for a Crossbonded Cable

A mean value for the series impedance matrix of one single major section is calculated as shown in Eq. 4.14.

$$\mathbf{Z} = \frac{1}{3}\sum_{i=1}^{3} \mathbf{Z}_{Section_i} \qquad [\Omega] \tag{4.14}$$

where $\mathbf{Z}_{Section_i}$ is the series-impedance matrix for one of the three minor sections.

Some new impedances are defined to describe the crossbonded cable system. The impedance $\underline{Z}_1$ is the self-impedance of the core, $\underline{Z}_2$ is the self-impedance of the sheath, $\underline{Z}_3$ is the mutual coupling between a core and sheath belonging to the same cable, $\underline{Z}_4$ is the mutual coupling between core/sheath between cable 1 and 2, $\underline{Z}_5$ is the mutual coupling between core/screen and core/screen between cable 1 and 3, and $\underline{Z}_6$ is the mutual coupling between core/screen and core/screen between cable 2 and 3. Defining $\mathbf{Z}_{section1}$, $\mathbf{Z}_{section2}$ and $\mathbf{Z}_{section3}$ based on the impedances just described and using Eq. 4.14, a matrix representing the mean one major section is obtained.

$$\mathbf{Z} = \frac{1}{3}\left[\begin{array}{ccc|ccc} 3Z_1 & 3Z_4 & 3Z_5 & Z_3+Z_4+Z_5 & Z_3+Z_4+Z_5 & Z_3+Z_4+Z_5 \\ 3Z_4 & 3Z_1 & 3Z_6 & Z_3+Z_4+Z_6 & Z_3+Z_4+Z_6 & Z_3+Z_4+Z_6 \\ 3Z_5 & 3Z_6 & 3Z_1 & Z_3+Z_5+Z_6 & Z_3+Z_5+Z_6 & Z_3+Z_5+Z_6 \\ \hline Z_3+Z_4+Z_5 & Z_3+Z_4+Z_6 & Z_3+Z_5+Z_6 & 3Z_2 & Z_4+Z_5+Z_6 & Z_4+Z_5+Z_6 \\ Z_3+Z_4+Z_5 & Z_3+Z_4+Z_6 & Z_3+Z_5+Z_6 & Z_4+Z_5+Z_6 & 3Z_2 & Z_4+Z_5+Z_6 \\ Z_3+Z_4+Z_5 & Z_3+Z_4+Z_6 & Z_3+Z_5+Z_6 & Z_4+Z_5+Z_6 & Z_4+Z_5+Z_6 & 3Z_2 \end{array}\right] \tag{4.15}$$

Assuming a trefoil formation, the impedances $\underline{Z}_4$, $\underline{Z}_5$ and $\underline{Z}_6$ become equal. As a result the sub-matrices $\mathbf{Z}_C$, $\mathbf{Z}_S$ and $\mathbf{Z}_{CS}$ defined in Eq. 4.7 all become balanced. The sequence components describing one full major section can therefore be obtained without loss of accuracy. For a system in flat formation this is not the case as only $\underline{Z}_4$ and $\underline{Z}_6$ are equal and off-diagonal elements different from zero are present.

Assuming bonded and solidly grounded sheaths at the major section's ends, the positive sequence impedance with eliminated sheath can be calculated using Eqs. 4.12 and 4.13. For crossbonded cables, the sheaths are, however, not solidly grounded at major sections ends, but grounded through the local ground electrode in the field. As a result Eq. 4.13 will therefore only give an estimate of the sequence impedances both for systems in trefoil and flat formation.

The full series impedance matrix and the sequence impedance matrix with and without eliminated sheaths are only valid for calculating a mean value of the impedance for one full major section. Faults can occur anywhere along major sections and longer cable systems are constructed using more than one major section. Therefore, the fault loop impedance as a function of fault location is of interest in order to evaluate whether the mean value given in Eq. 4.15 is a good approximation for the fault loop impedance per phase and whether the representation can be used for fault location or not.

Fault impedance calculations in crossbonded systems are very complicated and an iterative solution method is often used. This allows a full representation of the zero-sequence system and it is an easy way to conduct sensitivity analysis on all parameters. In the next section, a complete steady state model of a two major section crossbonded cable is implemented in DIgSILENT Power factory and the behaviour of the fault loop impedance is examined.

4.3 Fault Loop Impedance on Crossbonded Cable Systems

In this section, the fault loop impedance of several crossbonded cable systems are determined based on voltages' and currents' phasors measured at the fault locator terminals. It is in all cases assumed that the fundamental power frequency component can be extracted based on the the transient fault signal with no error.

The goal of this section is to determine which parameters are of importance for the fault loop impedance and therefore important for an impedance-based fault location method. The single core to sheath fault is most common and the study is focused on this fault type.

The sequence components are as mentioned often used as input for impedance-based fault locator methods and are in most cases measured after cable installation. The fault loop impedance for a single phase fault is determined based on symmetrical components as shown in Eq. 4.16 [13].

$$\underline{Z}_F = \frac{2\underline{Z}_1 + \underline{Z}_0}{3} \cdot l \qquad [\Omega] \tag{4.16}$$

where $\underline{Z}_1$ is the positive sequence impedance and $\underline{Z}_0$ is the zero sequence impedance of the cable system.

Use of the sequence components can provide a measurement of the expected fault loop impedance assuming a linear relationship between the fault location and the measured impedance. As already discussed, the impedance based on symmetrical components is only valid for an ideally transposed cable system with no crossbondings. The simple linear calculation method is in many cases a good approximation for OHL systems and widely used for fault location on such systems. This is, however, not the case on crossbonded cable systems due to the discontinuities at the crossbonding points [8, 15, 16]. To examine the behaviour in further detail and relate it to fault location, a model of a 12 km two major section cable system with minor sections of 2 km is implemented in DIgSILENT Power Factory. DIgSILENT Power Factory offers the possibility to implement full models of crossbonded cables with use of state-of-the-art cable models and a correct representation of the sheath sequence system. The case study 165 kV single-core cable laid in flat formation described in Sect. 4.1 is used and the setup implemented in DIgSILENT is shown in Fig. 4.3.

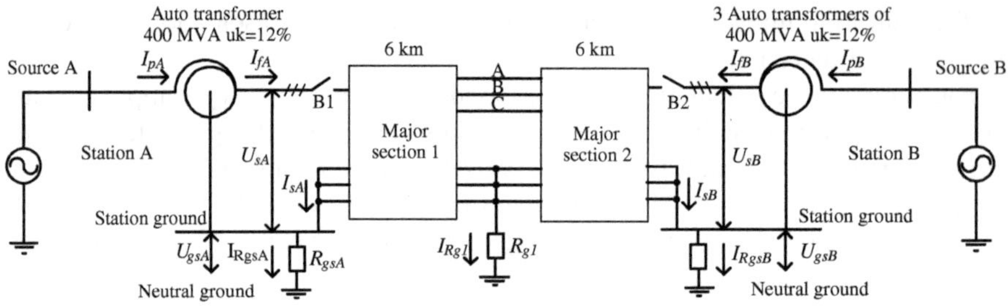

Fig. 4.3 Case study system including substation A and B and a two major section crossbonded cable

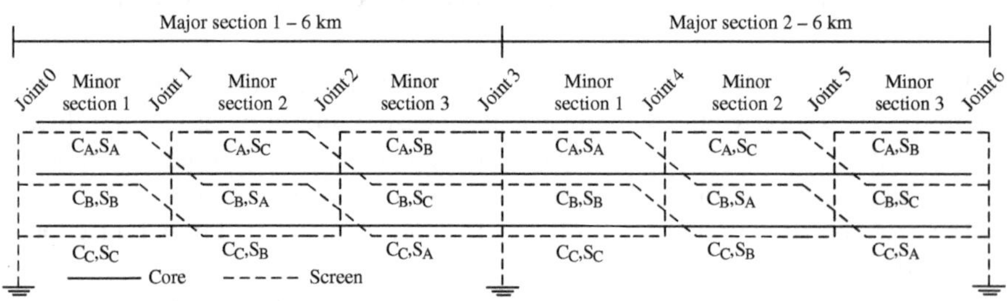

Fig. 4.4 Cable crossbonding scheme

The cable model takes a physical description of the cable as input. The thickness of each layer and their electrical parameters are given, and, based on these, the series impedance and shunt admittance matrix are formulated. A distributed line model is used for each minor section. The model's parameters were all presented in Table 4.2. The crossbonding scheme is shown in Fig. 4.4.

The breakers B1 and B2 in Figure 4.3 disconnect the cable system from the respective station and in general, B2 is open so the cable system is only fed from Station A. The equivalent resistance between the substations grounding grid and neutral ground at Station A and B are modeled as R_{gsA} and R_{gsB} and are set to 0.1 Ω when both B1 and B2 are closed. If B2 is open, R_{gsB} is modelled as a field grounding resistance and is together with R_{g1} set to 5 Ω. The cable sheaths are bonded and connected directly at the substation's grounding grid. All impedances between the connection point of the cables and the auto transformer are neglected. A single 400 MVA auto transformer with a short circuit voltage of 12 % is used at Station A and three 400 MVA auto transformers feeding in on Terminal B with the same short circuit rating are used at Station B. All transformers are connected to an infinite grid. The influence of the different short circuit power at Station A and B is important for double-sided infeed studied later.

Using these parameters, the positive and zero sequence impedances are determined for the cable as $\underline{Z}_1 = 0.2132\angle 82.7\,\Omega/\text{km}$ and $\underline{Z}_0 = 0.1243\angle 30.70\,\Omega/\text{km}$ measured from Station A. An ideal fault ($R_f = 0$) is applied between core and sheath on the three phases one at the time while the fault is moved from Station A

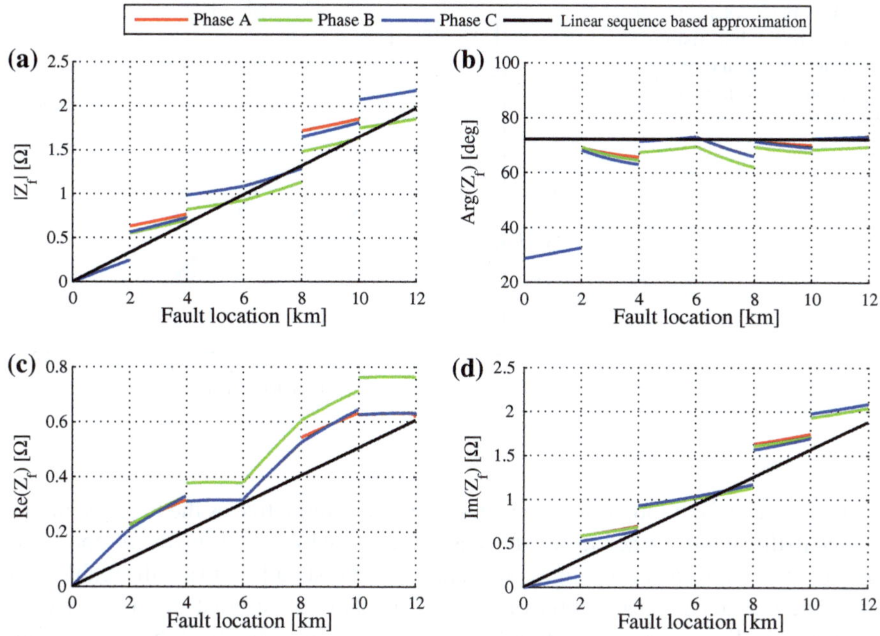

Fig. 4.5 Fault loops impedance for a single core to sheath fault on a 12 km crossbonded 165 kV cable laid in flat formation

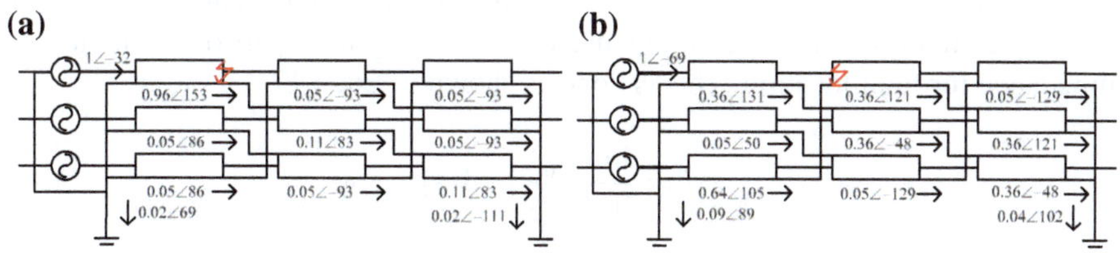

Fig. 4.6 Sheath current flow for a single phase to sheath fault at **a** the end of the first minor section and **b** at the beginning of the second minor section

towards Station B. The magnitude, the angle, the real and imaginary part of the fault loop impedance measured at Station A are presented in Fig. 4.5 together with the linear approximation of the fault loop impedance-based on Eq. 4.16.

The discontinuities in the fault loop impedance are caused by the sudden change in the return path of the fault current when the fault shifts from one side of a crossbonding to the next. In Fig. 4.6, the resulting per unit sheath currents are presented for a core-sheath fault at the end of the first minor section and for the same fault type at the start of minor section two. All angles are given relative to the voltage of phase A.

The very different flow of return currents for a fault at almost the same location gives rise to a discontinuity in the fault loop impedance at the crossbondings (2, 4,

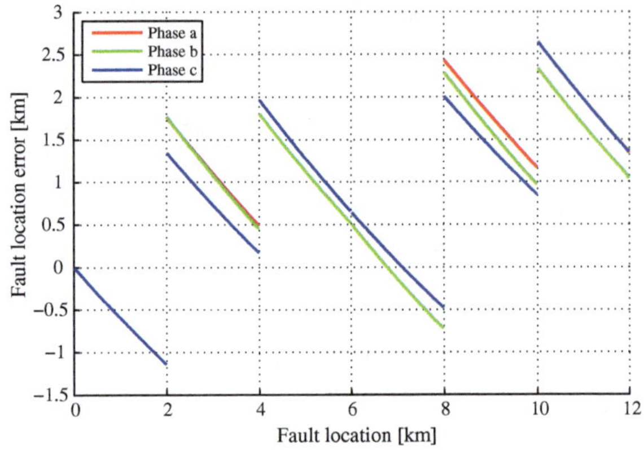

Fig. 4.7 Fault location estimation error when using a simple imaginary representation of the fault loop impedance

6 km, etc). The fault loop impedance becomes more inductive as the fault location is moved to the second minor section because the current loop made up by the conductor and combined return path changes physical size as the sheath from another cable must carry the major part of the return current. It is also noticed that the impedance of one phase is different compared to the impedance of the other two phases. The fault location estimation error is therefore also dependent on in which phase the fault occurs.

As mentioned in the review of the current fault location methods, several of the simpler fault location algorithms use only the imaginary part of the fault loop impedance for fault location. This is done as shown in Eq. 4.17.

$$d = \frac{\mathrm{Im}(\underline{Z}_{\mathrm{fl}})}{\mathrm{Im}(\underline{Z}_{\mathrm{L}})} \qquad [\mathrm{m}] \tag{4.17}$$

where d is the distance to fault, $\underline{Z}_{fl}$ is the measured fault loop impedance and $\underline{Z}_L$ is the line impedance. Even though the method is simple, it will provide an almost exact estimate of the fault location on transposed OHLs if all line parameters are known. Figure 4.7 shows the fault location error on the 12 km crossbonded cable using this simple approach.

Very large errors are introduced by assuming a linear dependent fault loop impedance for crossbonded cables. Furthermore, the different in fault loop impedance between the phases causes different errors for faults at the same location on the different cables. As expected, the simple fault location methods are not applicable for crossbonded cable systems.

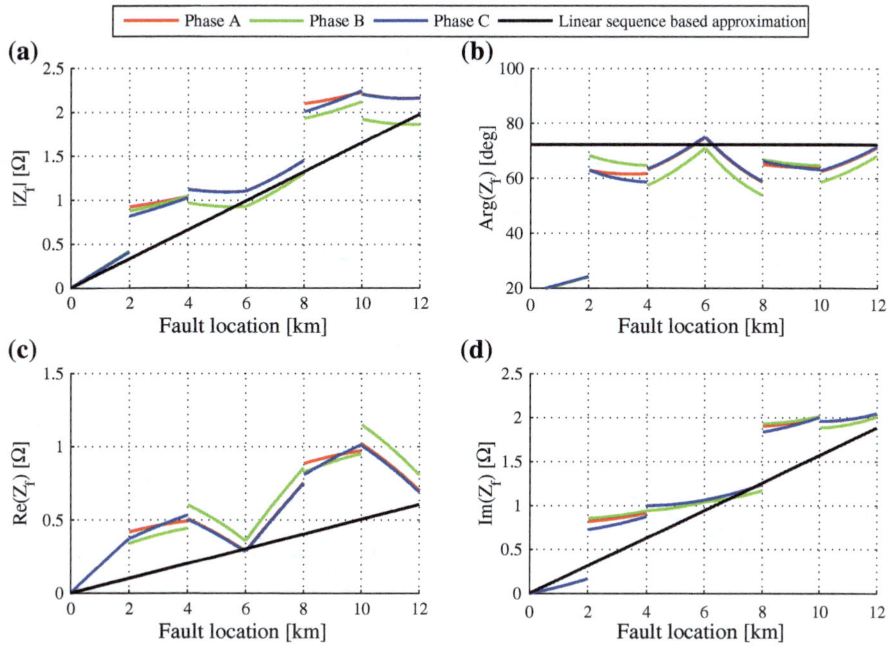

Fig. 4.8 Fault loops impedance for a single core to sheath fault on a 12 km crossbonded 165 kV cable laid in flat formation with double-sided infeed

4.3.1 Double-sided Infeed

Figure 4.8 shows the fault loop impedance if breaker B2 in Fig. 4.3 is closed (double-sided infeed) while a single core to sheath fault is moved from Station A to B. The fault resistance is set to zero and no connection to ground is present at the fault location (outer jacket is not damaged).

For single phase bolted faults on an OHL-system, the fault loop impedance is not affected by double-sided infeed [17]. If Fig. 4.8c and d are compared to Fig. 4.5c and d, it can be seen that this does not hold true for crossbonded cable systems. The voltage drop in the fault loop is not only caused by the current measured at the fault locator terminal, but also by the current from the far-end source. The cable's sheaths carries return current from both sources flowing dependent on the relationship between the impedance provided by the return system. The effect on the reactive part of the impedance can either be inductive or capacitive dependent on the relation between the angles of the currents at both terminals. This effect is similar to the 'Reactance effect' known from distance protection and is a further complication for accurate fault location on crossbonded cable systems. This effect must be included if an impedance-based fault location method is chosen.

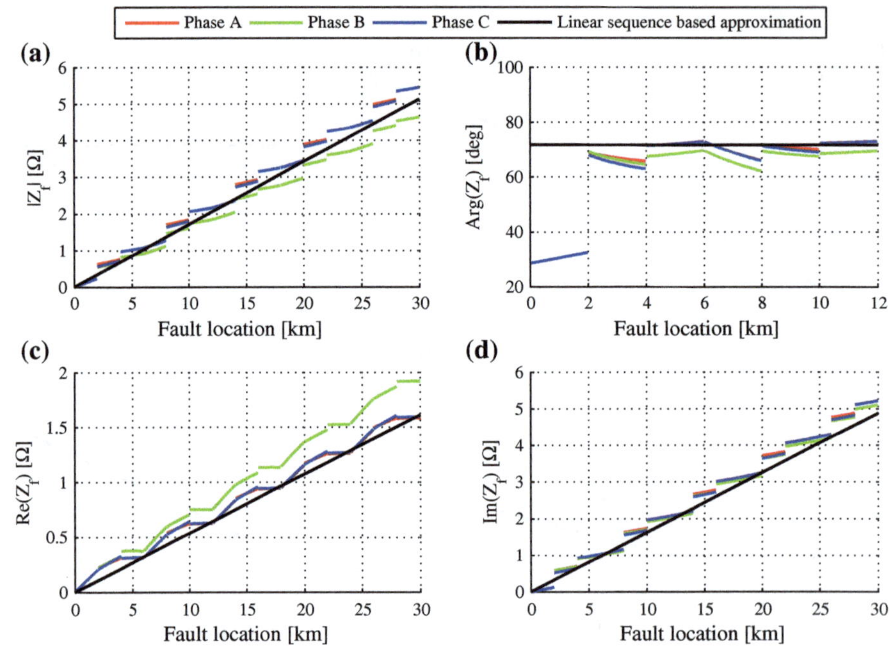

Fig. 4.9 Fault loop impedance for a single core to sheath fault on a 30 km crossbonded 165 kV cable laid in flat formation

4.3.2 Long Cables

The discrete changes in the fault loop impedance arise due to the change in the return path for the fault current. The relative change is larger for shorter cables where it is expected that the influence is to some degree averaged out on longer systems. The case study 165 kV cable is used to construct a 5 major sections cable with minor section of 2 km as well. This makes the total cable length 30 km. The fault loop impedance for the three phases is re-calculated for single-sided infeed. The result is shown in Fig. 4.9.

As expected, the relative error is smaller for longer cables. Figure 4.10 shows the resulting error if the reactance method is used to estimate the fault location.

Even though the relative error is reduced, the absolute error remains the same. The absolute error is the true parameter on which a fault location method should be evaluated as this error will determine the amount of digging work that has to be carried out.

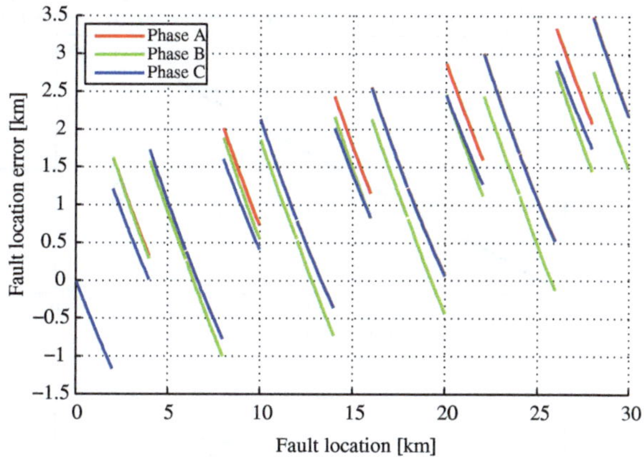

Fig. 4.10 Fault location estimation error for using the reactance method on a 30 km crossbonded cable system

4.3.3 Trefoil Formation

A cable system laid in trefoil will have equal mutual impedances due to symmetry wherefore symmetrical components can express the mean value of the cable's impedance with no error. Furthermore, the cables are laid closer together compared to a system in flat formation which will reduce the effect of the discrete changes in the impedance. The 165 kV case study system is placed in touching trefoil and the fault loop impedance calculated per phase. The result is shown in Fig. 4.11.

As expected, the fault loop impedance between the phases becomes equal due to equal mutual coupling. Discontinuities are still observed as the size of the current loop still changes abruptly when the sheaths are transposed. This is still a problem unique for crossbonded cables.

4.3.4 Fault Loop Impedance as Function of Cable and Cable System Parameters

It is quite well known how different materials and their electrical properties affect the impedance and admittance matrices of a three single-core high voltage cable system. The electrical properties of the materials are not expected to chance much over time once the cable system is laid; only changes in the resistive part of the fault impedance due to temperature variations can occur. However, system parameters such as the ground resistivity, grounding resistance in the field and at the substation will influence the complex return path of the fault current and thereby the impedance measured at the cable's terminals. In the following section, the influence on the fault loop impedance of these parameters is examined.

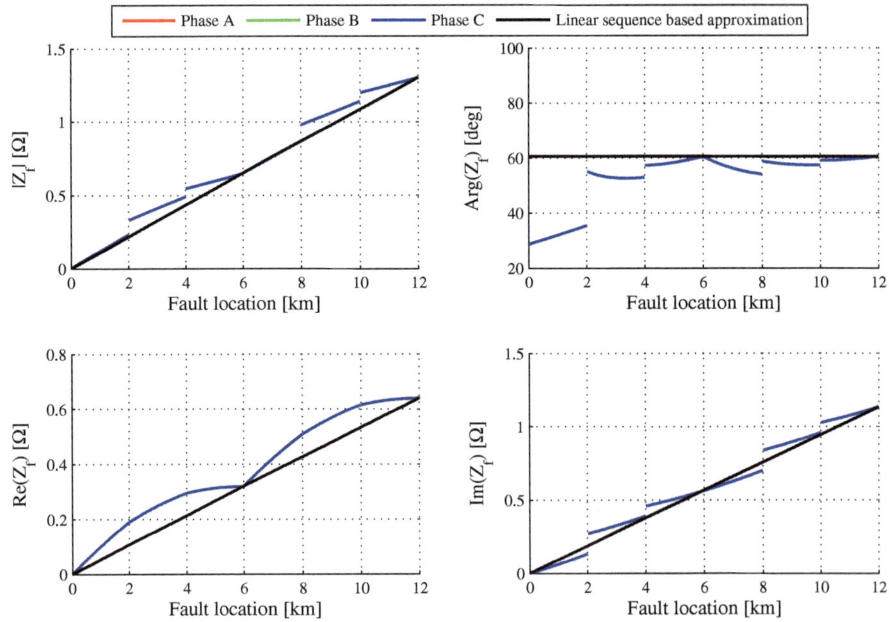

Fig. 4.11 Phase fault loops impedance for a single core to sheath fault on a 12 km crossbonded 165 kV cable laid in touching trefoil

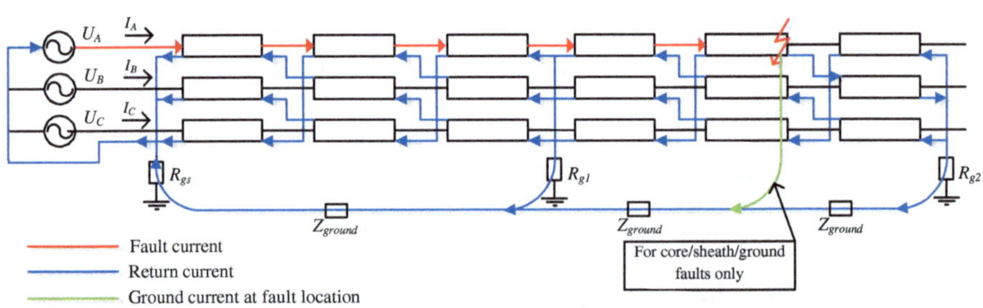

Fig. 4.12 Fault and return current in case of a short circuit in the second minor section of the second major section of a crossbonded cable with single-sided infeed

Field- and Substation Grounding Resistances and Ground Resistivity

Figure 4.12 shows the flow of fault and return current in case of a short circuit in the second minor section of the second major section on a crossbonded cable with single-sided infeed. The figure shows as well the additional ground current flowing if the outer jacket is damaged and contact to the surrounding soil is established (green line).

The distributed ground impedance is in Fig. 4.12 included as Z_{ground}. The ground impedance is dependent on the ground resistivity which can change locally and during the year. Furthermore, the equivalent grounding resistance at the substation (R_{gs}) and

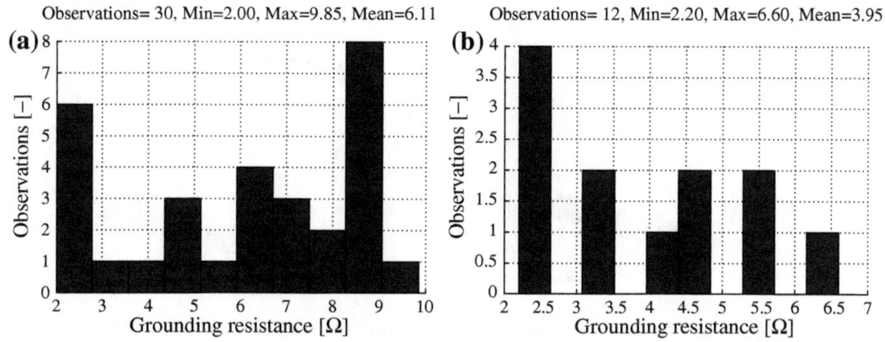

Fig. 4.13 Grounding resistance for **a** the cable line Blåvand-Karlsgårde and **b** the cable line Ådalen-Gistrup

in the field (R_{g1} and R_{g2}) will contribute to the overall impedance of the equivalent return path, thus, variations in these can affect the fault loop impedance as well. The requirements for the value of the grounding resistance are according to Cigré that resistance should below 20 Ω [18]. Energinet.dk has a more strict demand of a field grounding resistance below 10 Ω. The equivalent field grounding resistance is measured when a new cable line is installed in Denmark. Histograms of such measurements from the Danish cable lines Blåvand-Karlsgårde and Ådalen-Gistrup are presented in Fig. 4.13a and b.

As seen in the histograms, the resistance varies substantially even for the same cable line and the effect must be examined.

The part of the fault current flowing in the ground has to return to the source through the grounding resistance locally at the substation. This resistance is kept very low, but as the fault loop impedance of the cable system is relatively low as well, it can have some influence especially for faults close to the station. The magnitude of the fault loop impedance measured on the case study 165 kV cable system is plotted in Fig. 4.14a against the fault location and the field grounding resistance varied from 0 to 10 Ω (R_{g1} and R_{gsB} in Fig. 4.12). The substation grounding resistance R_{gsA} is set to 0 Ω as a worst case study.

In Fig. 4.14b, the magnitude of the fault loop impedance is plotted against the fault location and the grounding resistance at the substation (R_{gsA} in Fig. 4.12). The resistance is varied from are kept constant at 0 Ω as a worst case study.

Figure 4.14a and b show that the measured fault loop impedance is almost independent of the station and field grounding resistances. This is due to the high impedance return path provided by the ground compared to the sheaths. The sheaths are normally connected directly at the substation's grounding grid, thus most return current is flowing in the cable's sheaths for realistic fault on cable systems. This is, however, under the assumption that no metal is present near the cable system. If a metallic path parallel to the system (other cable systems, water pipes, rail road tracks, etc) is present, the zero sequence impedance becomes very difficult to predict and location dependent.

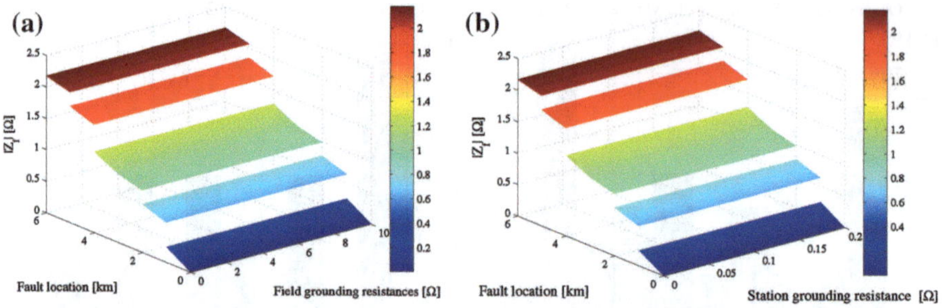

Fig. 4.14 Magnitude of fault loop impedance as function of fault location and **a** field grounding resistance and **b** substation grounding resistance

From the study of the grounding resistances, it can be concluded that the fault loop impedance is quite independent of the soil in which the cable system is laid as the large inductance of the ground loop combined with the grounding resistances limit the currents returning in the ground. This is a benefit for fault location purposes as the cable and its material are well known contrary to the soil properties.

Earth resistivity is dependent on several factors; soil temperature, water content and type of ground. The water content and the presence of other metallic conductors are two of the main parameters determining the earth resistivity. In urban areas such as Copenhagen, the presence of other metallic conductors is high. However, the fault current must return close to the fault carrying conductor due to AC-properties and only the very close metallic conductors will influence the fault loop impedance. As the ground resistivity decreases, more current returns through the ground and the effect of the ground and the grounding resistances become of greater importance. Figure 4.15 shows the magnitude of the fault loop impedance as the ground resistance is varied from 5 to 280 Ωm. The station and field grounding resistances are set to 0 Ω to study the full effect of the ground resistivity.

From Fig. 4.15, it can be concluded that even for a low ground resistivity, very limited current will return to the source through the ground. In the case presented, the change in impedance between a ground resistivity of 5 and 280 Ω is 2.0 %. This means that local variations in the ground resistance are of less importance as long as no parallel metallic parts are present. If this is the case, the system changes dramatically due to the relatively much lower impedance path introduced.

Core to Sheath to Ground Faults

In some cases, the outer jacket of the single core cable can be damaged in such a way that a connection to neutral ground is established. A part of the fault current can return to the source from this point through the ground, thus, an effect on the fault loop impedance measured at the fault locator terminals can be seen.

Fig. 4.15 Magnitude of fault loop impedance as function of fault location and ground resistivity

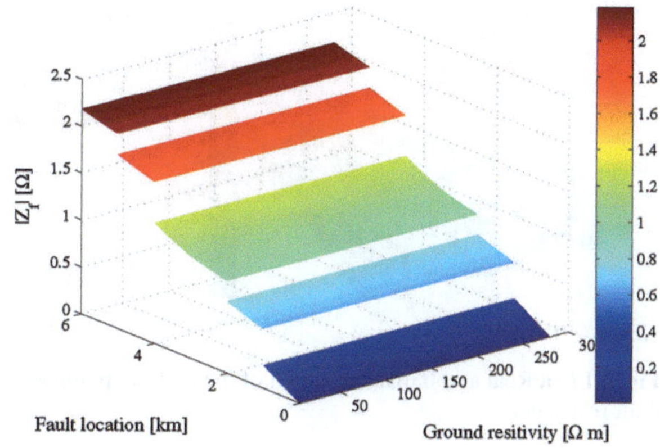

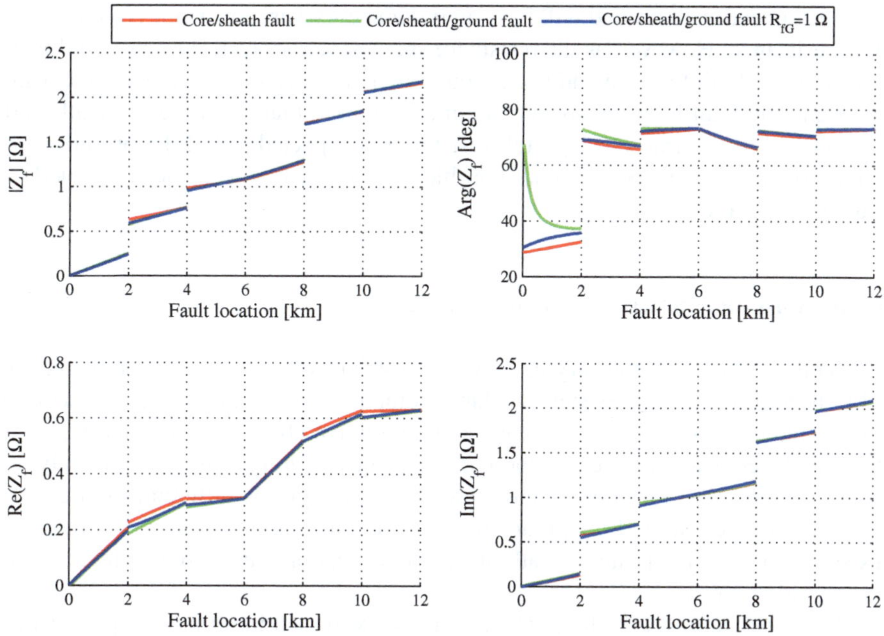

Fig. 4.16 Fault loops impedance for a single core to sheath fault and single core to sheath fault with a connection to neutral ground of 0 and 1 Ω on a 12 km crossbonded 165 kV cable laid in flat formation

The equivalent resistance to neutral ground at the fault location is very difficult to model as it will depend on the local soil parameters and the extend and type of damage to the outer jacket. In Fig. 4.16, simulations are presented showing the effect of a direct connection between the sheath and neutral ground ($R_{fg} = 0$) and if an equivalent resistance of 1 Ω represents the connection.

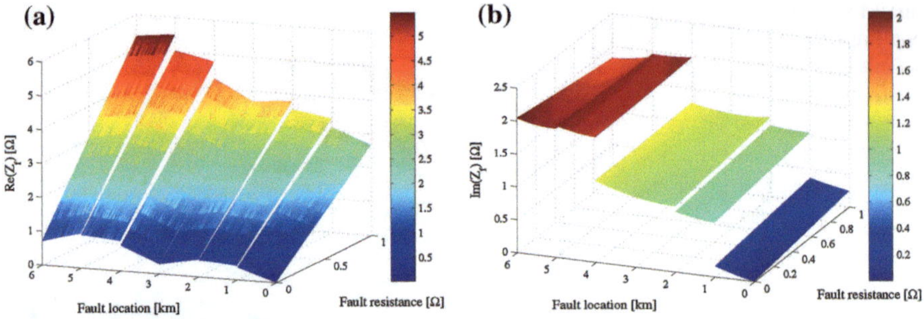

Fig. 4.17 **a** Real and **b** imaginary part of the fault loop impedance as function of fault location and fault resistance

The connection to ground will have an influence for very low values of the equivalent resistance. However, as the resistance increases, very little current will flow to ground at the fault location and for a grounding resistance of 1 Ω, almost no current flows and the fault loop impedance is almost equal to that of a core to sheath fault only. In real life, a time varying fault resistance is expected [19, 20]. For fault location problems, this can be a potential problem, as a very high accuracy is needed for exact fault location.

Fault Resistance Between Core and Sheath

The fault resistance is time varying and can vary from close to zero to some ohms on cable systems [21]. This resistance is large compared to the impedance of especially short cable systems and can thus have a large relative effect. The imaginary part of the impedance is not affected by the fault resistance as long as the cable is only fed from a single side - for double-sided infeed, the situation is different. Figure 4.17a and b shows the real and imaginary parts of the fault loop impedance as the fault resistance between core and sheath at the fault location varies from 0 to 1 Ω with double-sided infeed.

The influence of the fault resistance can be expressed in terms of a phasor representation of the currents measured at Station A and B and the real fault resistance R_F as shown in Eq. 4.18 [17].

$$\underline{R}'_F = R_F \frac{\underline{I}'_A + \underline{I}'_B}{\underline{I}'_A} \tag{4.18}$$

where R_F is the real fault resistance, $\underline{I}'_A$ and $\underline{I}'_B$ are the fault currents measured at Station A and B respectively.

Equation 4.18 shows that the influence of the fault resistance measured from Station A will dependent on the contribution of currets from both cable terminals. Depending on the difference in phase between the two source currents, the fault resistance R_F becomes a complex number. The results is a change in both the real and imaginary parts of the fault loop impedance measured from the terminal. The magnitudes of the current will determined the influence of the fault resistance. If I'_A is dominating (strong source), the influence is reduced where it for a weak source in the measuring end will lead to a magnification of the fault resistance seen from the fault locator point.

As shown in Fig. 4.17, for a fault resistances of 1 Ω and a fault location 0.2 km from Station A, the fault loop impedance is determined as $2.66 + j0.24\,\Omega$ where for single-sided infeed with a fault at the same location, it is approximately $1 + j0\,\Omega$ (see for instance Fig. 4.5). The current feed into the fault from Station A is $10.9\angle 107.8$ A and $18.3\angle 116.15$ A from Station B. Using Eq. 4.18 and a fault resistance of 1 Ω, an increase in the measured fault loop impedance of $2.66 + j0.24\ \Omega$ is found. This fully explains the large impedance measured at Station A.

The reactance effect is especially important for cable systems even with small fault resistances as the fault loop impedance per unit length is small compared to OHLs. An efficient way to compensate for the influence is hence necessary for an impedance-based fault location method for crossbonded cables.

Section Summary

The fault loop impedance shows discrete changes at every crossbonding. This is due to the change in return path for the fault current. Double-sided infeed has an affect on the imaginary part of the fault loop impedance for bolted faults on crossbonded systems due to the common return path for the two sources provided by the sheath system.

On longer cables, the effect of the discrete changes becomes less relatively seen, but the absolute fault location error using a simple reactive method is the same. For a cable system in trefoil, the fault loop impedance per phase becomes equal, but discrete changes are still observed.

The fault loop impedance is not much affected by the equivalent resistance of the grounding rods in the field or by the grounding of the substation due to the large ground impedance. Therefore, if the cable model predicts the fault loop impedance well based on cable parameters, it may be possible to develop an impedance-based fault location method for crossbonded cables. However, for double-sided infeed, a reactance effect is seen in the measured impedance especially for far-end faults. The effect is dominating and a compensation schemes must be implemented. In the following section, impedance-based fault location on hybrid systems is examined.

4.4 Fault Location on Hybrid Lines Using Impedance-Based Methods

Line configurations different from the classical crossbonded cable system are planned in the future Danish transmission grid. Overhead lines and cables connected in a radial formation (hybrid lines) will form a 400 kV backbone line interconnecting the main Danish sub stations. Furthermore, during the transition period from an OHL to a cable based network, some hybrid lines may be in operation. Transmission lines connecting offshore wind farms to the main grid already exist in the Danish network and several new ones are in the planning stage. Typically, a crossbonded land cable constitutes the first part of the transmission line from a suitable substation to the coast wherefrom a submarine cable is used. Such a line can in some case only be monitored from the substation on land and from the offshore platform and is therefore a hybrid line. A fault location method for such a system is very important as extended offshore fault location time can be very costly. In this section, the use of an impedance-based fault location method for hybrid lines is examined.

4.4.1 *The Fault Loop Impedance of a Hybrid Line*

The fault loop impedance of OHLs and crossbonded cable systems is very different. The small spacing between the conductors on cables makes the reactance per unit length small compared to OHL-system. Typical values for the fault loop impedance are 0.1 Ω/km for cable in touching trefoil and 0.16 Ω/km for systems in flat formation where for OHL-systems, the reactance is around 0.6 Ω/km. Furthermore, the shunt capacitance per unit length for cables is much larger than for OHLs due to the use of an insulation material with a relative permittivity of more than one and because of the small spacing between core and sheath.

Most of the faults that occur on OHLs result from isolator flashover. The arc resistance is therefore in series with the tower equivalent impedance which again is in series with the tower footing resistance. From the tower, the current can return to the source through the ground. If ground wires are used, a part of the current will flow in these as well. The ground wires are solidly connected to all towers, so a return path is also provided by the tower and footing resistance of the other towers. How the current is divided is determined by local parameters such as ground resistivity and the individual tower footing resistances [21].

For hybrid lines, the return path will include the cable's sheath circuit. Depending on the configuration of the line and in which segment the fault occurs (cable vs. OHL), the zero-sequence impedance will be very different. Normally, the sheaths are connected directly to the first tower's ground wires where a low impedance path is formed. The ground wires are, however, most likely made from steel which has a high resistivity compared to the materials used for the cable's sheaths. Furthermore, the average larger distance between the OHL conductors and the ground wires gives

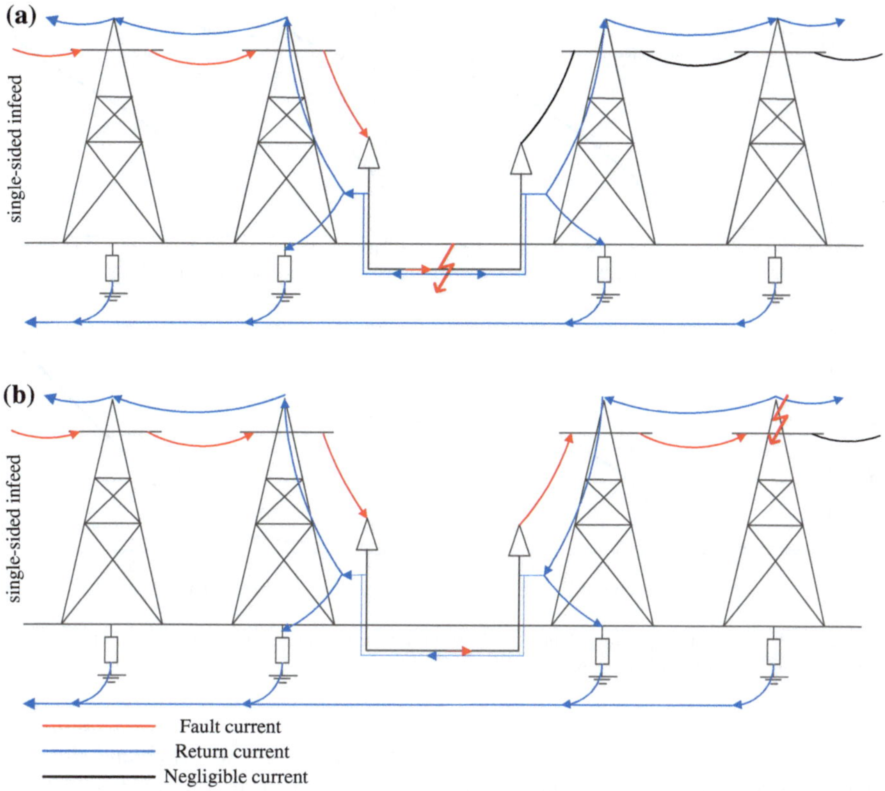

Fig. 4.18 Flow of current in case of a fault in a hybrid line in the **a** cable section and **b** in the OHL section (inspiration from [22])

rise to a larger reactive part. Figure 4.18 shows the flow of current for faults on a hybrid line where in (a), the fault occurs in the cable segment and in (b), the fault occurs on the far-side OHL segment.

Figure 4.18 shows the complicated and very different return paths for the currents in case of a fault in the cable or OHL-system. Due to the system structure of hybrid lines, the ground is normally more involved as return path due to the fact that the cable's sheaths can only carry the fault return current a part of the way back to the source.

A hybrid line is implemented in DIgSILENT to examine the behaviour of the fault loop impedance. The example hybrid line consists of the 165 kV case study cable (Sect. 4.1) and a 165 kV OHL line section which parameters are presented in Appendix B.1.3. The cable part consists of two major sections with minor sections of 1 km each. At the station where the cable and OHL are interconnected, the cable's sheath and the ground wirers of the first OHL-tower are both bonded and connected to the substation's grounding grid. It is assumed that the impedance between the cable's sheaths and the connection point of the ground wires is negligibly small.

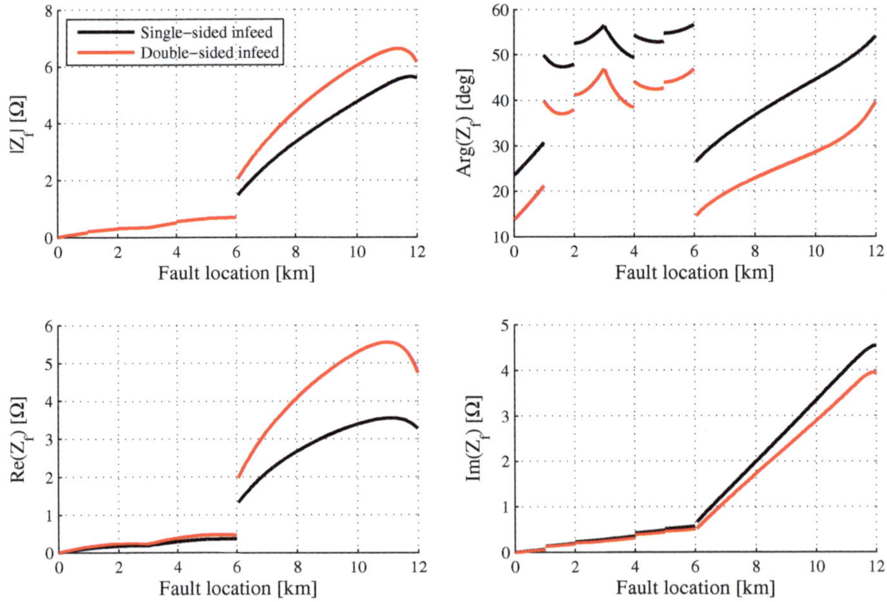

Fig. 4.19 Phase fault loops impedance for a single phase to sheath fault on a 30 km crossbonded 165 kV cable laid in flat formation

The grounding resistance of the cable substation is set to 0.1 Ω and the field grounding resistances in the cable system to 5 Ω. The tower footing resistance is modelled as a single resistance with a value of 1.5 Ω according to the methods described in [21].

The hybrid line is analysed when fed from both a single- and double-side while a single phase to sheath fault is applied on phase A along the cable section and then between phase A and the tower on the OHL-section. The arc resistance for faults in the OHL section is modelled as a constant resistance of 1 Ω placed between the conductor and ground wires. The fault loop impedance is shown in Fig. 4.19.

Figure 4.19 shows why hybrid lines are hard to monitor for fault location purposes using impedance-based methods. The different fault loop impedance of the two system types makes accurate estimation of a fault location difficult, especially on the cable section. Both the real and imaginary parts of the fault loop impedance are affected by double-sided infeed as for pure crossbonded cable sections which is a further complication.

4.5 Conclusions on the Fault Loop Impedance on Crossbonded Cable Systems for Fault Location Purposes

This chapter studies the behaviour of the fault loop impedance on crossbonded cables. It is found that the modelling of the zero-sequence system is very important for accurate results and a detailed representation of the sheath system is necessary. However, all parameters that mainly contribute to the fault loop impedance are parameters specific for the cables and their laying configuration where the parameters describing the ground and grounding resistances are of less importance. If the dominating parameters are well known and if a detailed model including the crossbonding of the sheath can provide good results when compared to real life, several of the fault location algorithms based on artificial intelligence can be trained to perform accurate fault location on crossbonded cables. If the models are not capable of this, the artificial neural network cannot produce good results when compared to real-life fault impedances.

It is evaluated that a practice implementation of an impedance-based fault location method for hybrid lines is not possible to implement if high accuracy is demanded. The different line parameters make the fault loop impedance of the different line types very different, thus, small errors in the overhead line parameters give large deviations in the faults loop impedance for fault in the cable sections.

References

1. S.A. Schelkunoff, The electromagnetic theory of coaxial transmission lines and cylindrical-shields. Bell Syst. Tech. J. **13**, 532–579 (1934)
2. F. Pollaczek, Uber das feld einer unendlich langen wechsel stromdurchflossenen einfachtleitung. Elektrische Nachrichten Technik **3**(9), 339–359 (1926)
3. L.M. Wedepohl, D.J. Wilcox, Transient analysis of underground power-transmission systems. System-model and wave-propagation characteristics. Proc. Inst. Electr. Eng. **120**(2), 253–260 (1973)
4. O. Saad, G. Gaba, M. Giroux, A closed-form approximation for ground return impedance of underground cables. IEEE Trans. Power Deliv. **11**(3), 1536–1545 (1996)
5. A. Ametani, A general formulation of impedance and admittance of cables. IEEE Trans. Power Apparatus Syst. **PAS-99**(3), 902–910 (1980)
6. IEC Standard 60 287-1-3, *Current Sharing Between Parallel Single-Core Cables and Calculation of Circulating Current Losses*, 1st edn (Geneva, Switzerland, IEC, 2002–2005)
7. C.L. Fortescue, Method of symmetrical co-ordinates applied to the solution of polyphase networks. AIEE Trans. **37**(Part II), 1027–1140 (1918)
8. T. Ohno, PhD-thesis, *Dynamic Study on the 400 kV 60 km Kyndbyvarrket Asnarsvarrket Line*, 1st edn. (Fredericia, Denmark, Energinet.dk, 2013)
9. U.S. Gudmundsdottir, PhD-thesis, *Modeling of Long High Voltage AC cables in Transmission Systems*, 2nd edn. (Fredericia, Denmark, Energinet.dk, 2010) ISBN 978-87-90707-73-6
10. B. Gustavsen, Panel session on data for modeling system transients insulated cables. in *Power Engineering Society Winter Meeting, 2001. IEEE*, vol. 2 (2001), pp. 718–723
11. B. Gustavsen, J.A. Martinez, D. Durbak, Parameter determination for modeling system transients-part ii: insulated cables. IEEE Trans. Power Deliv. **20**(3), 2045–2050 (2005)

12. C.F. Jensen, F. Faria da Silva, C.L. Bak, W. Wiechowski. Switching studies for the horns rev 2 wind farm main cable. in *International Conference on Power Systems Transients (IPST)*, (2011)
13. N. Tleis, *Power Systems Modelling and Fault Analysis—Theory and practice* (Elsevier, 2008) ISBN 13-978-0-7506-8074-5
14. Cigré Working Group B1.30, *Cable Systems Electrical Characteristics* (Paris, France, Cigre, 2013)
15. V. Leitloff, X. Bourgeat, G. Duboc. Setting constraints for distance protection on underground lines. in *Developments in Power System Protection, 2001, Seventh International Conference on (IEE)* (2001), pp. 467–470
16. D.A. Tziouvaras, Protection of high-voltage ac cables. in *Power Systems Conference: Advanced Metering, Protection, Control, Communication, and Distributed Resources, 2006. PS '06* (2006), pp. 316–328
17. M.M Saha, J. Izykowski, E.Rosolowski, *Fault Location on Power Networks* (Springer, 2010) ISBN 978-1-84882-886-5
18. Cigré Study Comittie 21, *The Design of Specially Bonded Cable Systems* (Paris, France, Electra N 28, Cigré, 2011)
19. B. Koch, Tests on xlpe-insulated cable arcing faults and arc-proofing. IEEE Trans. Power Deliv. **3**(4), 1289–1295 (1988)
20. M. Bashir, I. Niazy, J. Sadeh, M. Taghizadeh, Considering characteristics of arc on travelling wave fault location algorithm for the transmission lines without using line parameters. in *Environment and Electrical Engineering (EEEIC), 2011 10th International Conference on* (2011), pp. 1–5
21. G. Ziegler, *Numerical Distance Protection—Principles and Applications* (Publicis Kommunikation Agentur GmbH, GWA, Erlangen, 2008)
22. Cigré, *Earth Potential Rises in Specially Bonded Screen Systems* (Paris, France, Task Force B1.26, 2008), pp. 1–98

Chapter 5
Impedance-Based Field Measurements

In this chapter, impedance-based field measurements carried out on the Anholt cable are described. The task is to examine whether the most advanced cable models of today can be used to predict the fault loop impedance of various fault types is such a way that generic algorithms can be trained to estimate the fault location on crossbonded cables with good accuracy.

5.1 Anholt System Description

During the spring of 2012, the electrical link to the offshore wind park Anholt was being constructed. In the construction period, is was possible to perform field measurements on parts of the cable system.

The full cable system is divided into three parts. A 59.6 km land cable from the 400 kV substation Trige to a newly constructed shunt reactor station close to the beach by the Danish town Grenaa. The second part starts at the reactor station and ends at shore (0.5 km). From the shore to an offshore collector platform 24.5 km off the coast line, a three-phase submarine cable is used. In Fig. 5.1, a diagram of the electrical connection is shown.

The part of the cable system under study is the section of the land cable starting at substation Trige (Joint 0) till Joint 33 (38.32 km). At Joint 27, which is placed 31.415 km from Trige and 6.905 km from Joint 33, all cables were accessible to emulate various short circuits. A map showing the Anholt land cable is presented in Fig. 5.2.

The land cable system consists of three single core 2000 mm^2 milliken aluminum conductor cables as shown in Fig. 5.3.

The laying configuration is shown in Fig. 5.4.

The cables have a 170 mm^2 aluminum foil metal sheath and are placed in flat formation. An earth continuity conductor (ECC) is laid with the cable system to help reduce local earth potential rises caused by external fault currents flowing in the cable system [1].

C. F. Jensen, *Online Location of Faults on AC Cables in Underground Transmission Systems*, Springer Theses, DOI: 10.1007/978-3-319-05398-1_5,

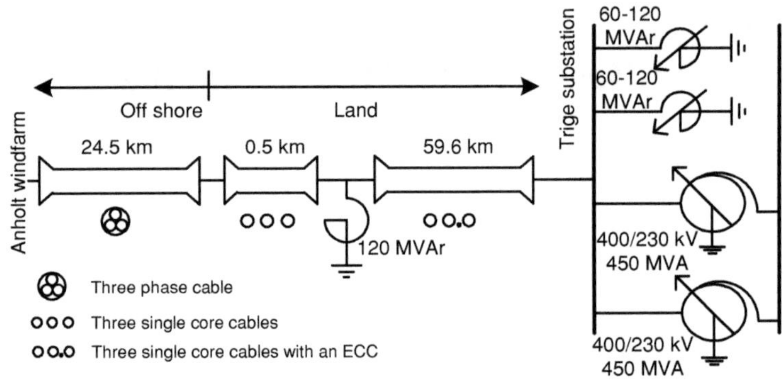

Fig. 5.1 Single line diagram of the electrical connection to the Danish offshore Wind Farm Anholt

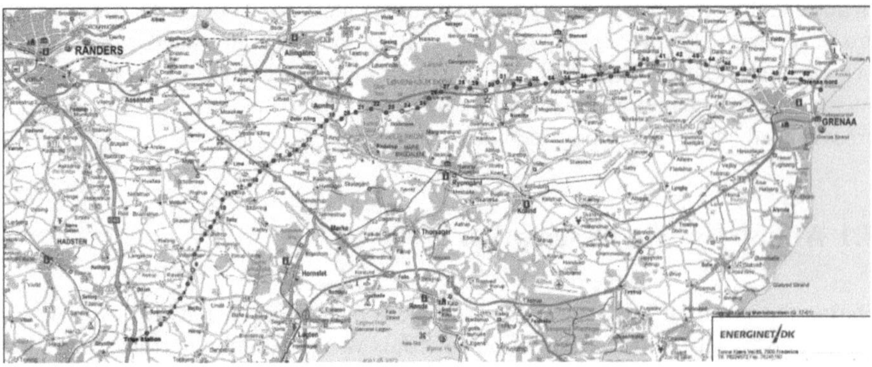

Fig. 5.2 Map showing the joints of the land cable for the Anholt offshore wind farm

Fig. 5.3 Physical layout of the 2000 mm^2 aluminium single core 245 kV cable

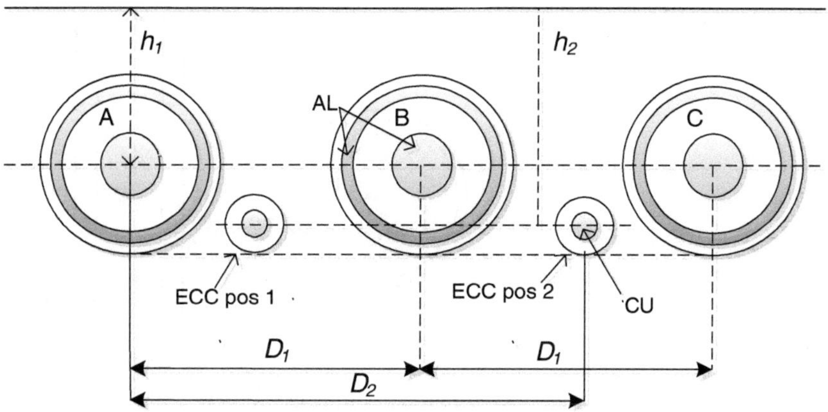

Fig. 5.4 Physical layout of the three single phase 2000 mm^2 aluminium single core 245 kV crossbonded cable system. $r_1 = 28.4$ mm, $r_2 = 53.9$ mm, $r_3 = 54.4$ mm, $r_4 = 58.6$ mm, $\rho_c = 3.547e^{-8}\,\Omega$m, $\rho_s = 2.676e^{-8}\,\Omega$m, $\rho_{ECC} = 1.724e^{-8}\,\Omega$m, $\rho_E = 100\,\Omega$m $\epsilon_{i1} = 2.89$, $\epsilon_{i2} = 2.3$, $\epsilon_{ECC} = 2.3$, $D_1 = 0.4$ m, $D_2 = 0.75$ m, $h_1 = 1.3$ m, $h_2 = 1.3511$ m

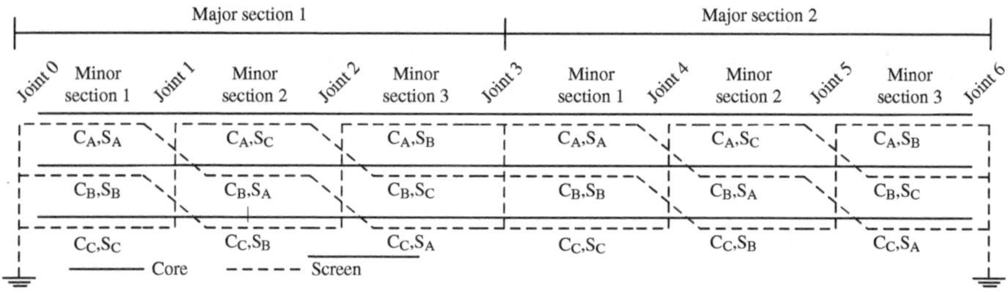

Fig. 5.5 Crossbonding method used for cable system part from Joint 0 to 27 with the ECC and surge voltage limiters excluded. C_i represents the core conductor of phase i and S_i represents the metal sheath conductor of cable i

The 31.4 km cable section is crossbonded and has nine major sections with the three metal sheaths bonded at the ends of the major sections, but grounded only at every second major section. In Fig. 5.2, all joints with bonded and grounded sheaths are marked in blue and all red markings indicate normal crossbonded joints. In Fig. 5.5, a sketch of the crossbonding method from Joint 0 to 6 is shown with the ECC and surge voltage limiters excluded.

As Fig. 5.2 shows, Joint 27 and 33 are not planned as joints with grounded sheaths, but to ensure a good measuring reference, it is, however, chosen to establish a temporary grounding system at these locations to which all cable sheaths and the measuring equipment can be connected. The cable section from Joint 27 to 33 is therefore normally crossbonded.

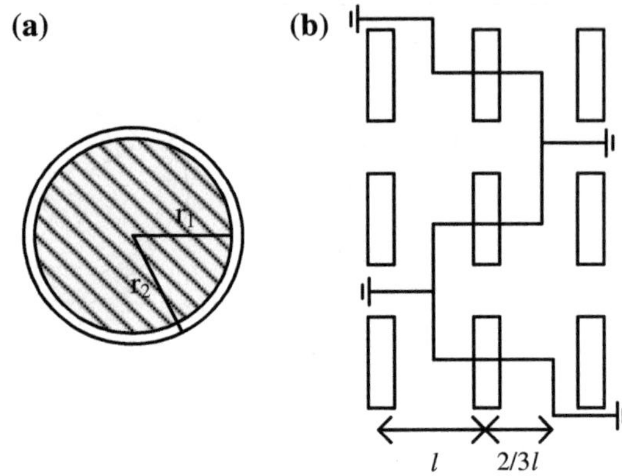

Fig. 5.6 **a** ECC construction and **b** Laying

5.1.1 Earth Continuity Conductor

An earth continuity conductor (ECC) consisting of an insulated 95 mm^2 copper cable is used in the Anholt system. The physical dimensions of the ECC is shown in Fig. 5.6a. The insulation material is PE with an relative permittivity of 2.3. The ECC is shifted as shown in Fig. 5.6b at the middle of each minor section.

5.2 Measuring Strategy

In this section, the measurement strategy for performing impedance-based field measurements on the Anholt cable is described.

The cable sections from Trige (Joint 0) to Joint 33 will as mentioned be available for measurements where Joint 27 is an open joint where all cable cores and screens can be accessed to make the desired short circuits. The cable system can also be seen as two cable systems (from Joint 0 to 27 and from Joint 27 to 33) where a fault can be applied at 100 % of the cable length. Summing up, the following six measuring case studies are possible and will be conducted. The fault location is given in brackets.

1. The sections from Joint 0 to 27 with faults at Joint 27 seen from Joint 0 (100 %).
2. The sections from Joint 27 to 33 with faults at Joint 27 seen from Joint 33 (100 %).
3. The sections from Joint 0 to 33 with faults at Joint 27 seen from Joint 0 (82 %).
4. The sections from Joint 0 to 33 with faults at Joint 27 seen from Joint 33 (18 %).
5. The sections from Joint 0 to 33 with faults at Joint 33 seen from Joint 0 (100 %).
6. The sections from Joint 0 to 33 with faults at Joint 0 seen from Joint 33 (100 %).

The measuring case studies are graphically presented in Fig. 5.7.

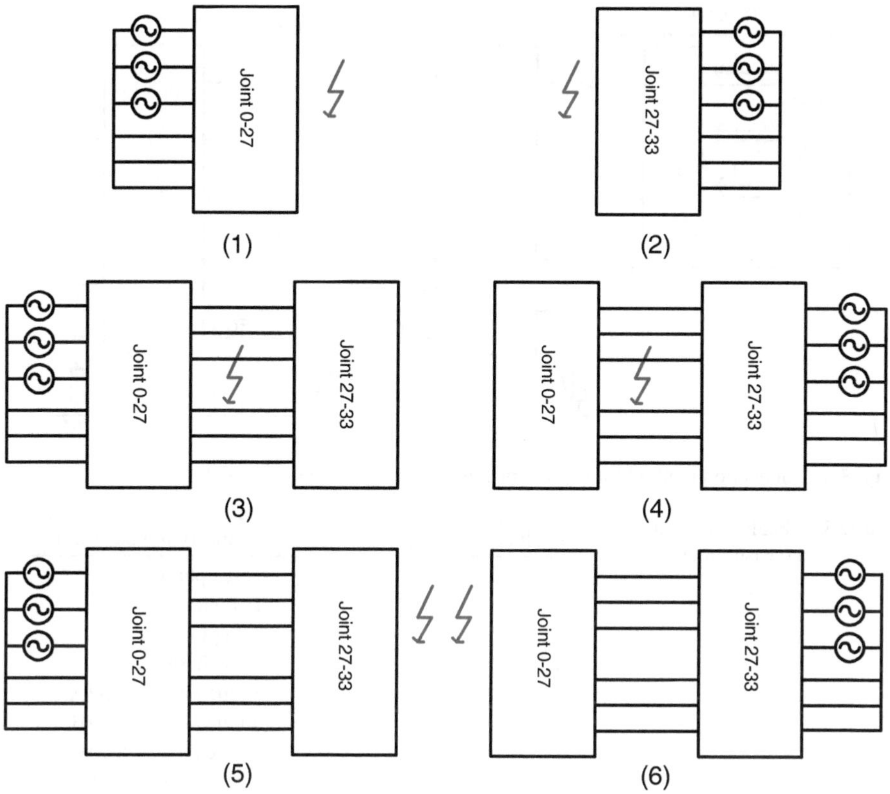

Fig. 5.7 The six different connection schemes for impedance measurements on the Anholt cable

A three-phase source will be used to supply the cable system while short circuits are applied at the selected locations. Only single sided infeed is tested as the task is only to determine the fault loop impedance of the cable system.

The voltage and current phasors are measured at the energised end, and from these, the fault impedance per phase can be calculated. Because the system is unsymmetrical in faulted conditions, it is of interest to do measurements on all three phases for all types of faults.

In order to limit the influence of the background noise, measurements will be performed at a range of frequencies around the power frequency; 30, 70, 90 and 110 Hz. The 50 Hz impedance can then be calculated based on interpolation assuming that the impedance is linear dependent on the frequency. The neutral of the three-phase source will be connected directly to the three bonded cable sheaths at the measuring location and to station ground or the local grounding electrode depending on the measuring location. The principle is shown in Fig. 5.8.

The different fault types are defined in Table 5.1.

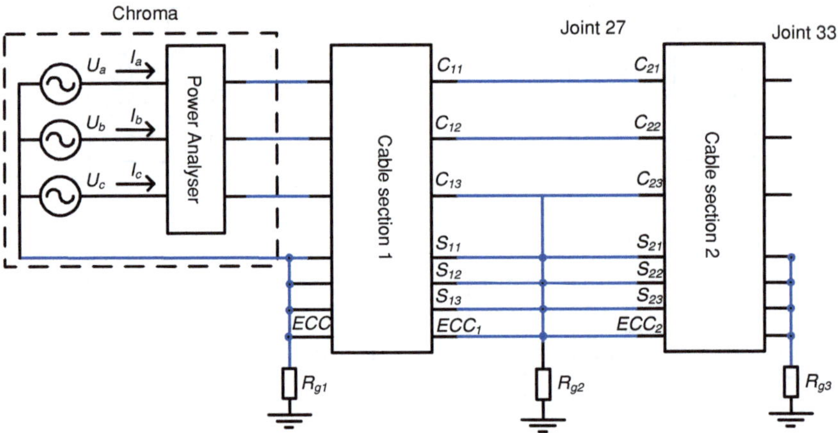

Fig. 5.8 Connection method for source for impedance-based measurements

Table 5.1 Fault scheme used to perform measurements

Fault type number	Connection type (phase)
1	Three phase
2	Two phase (AB)
3	Two phase (BC)
4	Two phase (CA)
5	Single phase CSG (A)
6	Single phase CSG (B)
7	Single phase CSG (C)
8	Positive sequence impedance
9	Zero sequence impedance

Table 5.2 Fault types applied during the six different connection schemes (see Table 5.1)

Connection scheme	Fault types
1	1–7
2	1–7
3	1–9
4	1–9
5	1–7
6	1–9

All faults involves the sheath of the faulted cable. Contact will always be established between core and sheath both for internal and joint faults. A three-phase fault is therefore in principle a three-core, to three-sheath to ground fault. Such a fault type will be very rare in real-life, especially for systems in flat formation but is included to study the models ability to determine the fault loop impedance for various faults. Table 5.2 states which short circuits are performed under which connection schemes. For cables sections where the fault location is at 100 %, the positive and zero-sequence impedances are measured as well.

Fig. 5.9 12 kVA Chroma 61511 power supply

Connecting the source at either Joint 0, Joint 27 or Joint 33 while applying the fault at one of the two additional locations makes measurements of a large set of short circuits possible.

5.2.1 Measuring Equipment

Chroma 61511

A three-phase power supply with variable frequency is needed. Available for the measurements is a 3×4 kVA Chroma 61511. The power supply is shown in Fig. 5.9.

The Chroma 61511 has a build in high precision power analyser which will be used to calculate the voltage and current phasors. The voltage and current accuracy is both $0.2\,\% + 0.2\,\%$F.S, and the measuring accuracy of the power analyser is $0.2\,\% + 0.2\,\%$F.S for voltage measurements and $0.4\,\% + 0.3\,\%$F.S (F.S = full scale). The full scale ratings are 150 V for the voltage and 32 A for the current.

Omicron CPC 100

To verify the positive and zero-sequence impedance measurements performed by the Chroma, the Omicron CPC 100 line impedance measurement device is used. The unit works with the CP-CU1 which is a multi-functional coupling unit for the CPC 100. The CPC 100 and the CP-CU1 are shown in Fig. 5.10.

These units measure the line impedance at different frequencies and then, the power frequency component is calculated by data treatment. The impedances are determined based on a seven-step measurement strategy. The method is described in Appendix C.

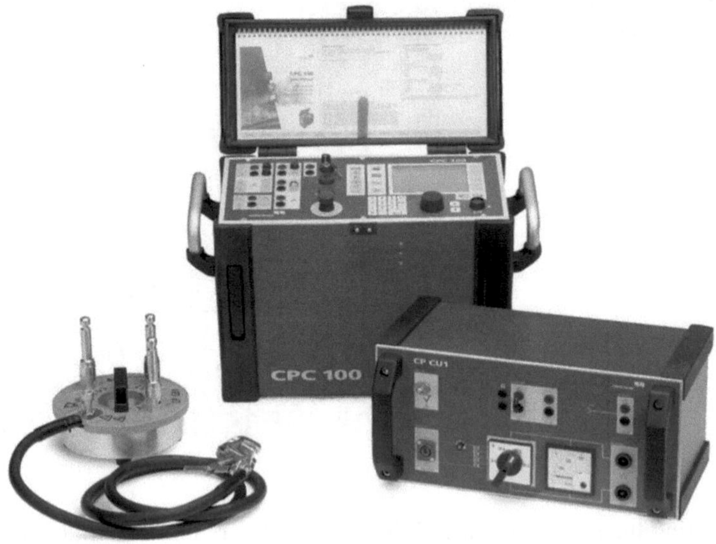

Fig. 5.10 CPC 100 line impedance measurement device

The seven-step measuring method assumes a balanced system and because the phase impedances Z_{1A}, Z_{1B} and Z_{1C} are determined based on inter-phase measurements only, the effects of the zero sequence system is not taken into account. For a crossbonded cable system in flat formation, due to the unbalanced phases, a small zero sequence current will flow when a positive sequence voltage is applied. The Chroma power source applies a positive sequence voltage as it is the case in a real-life power system, and the effect of the zero sequence current is therefore included in these measurements. The measurements performed using the Chroma are accurate in all conditions and the CPC 100 will only be used to measure positive and zero sequence impedances.

5.3 Performing Impedance-Based Measurements

The Chroma 61511, the CPC 100 and the CP-CU1 are installed in a car and generators are placed at Joint 27 and 33. All signal measurements are monitored by oscilloscopes to detect errors. Firstly, a fault is applied at the desired fault location by an instructed person. The Chroma will be connected at the measuring location and the output voltage will slowly be raised until 4 kVA is supplied to the phase with the peak load. The fault loop impedance is measured at 30, 70, 90 and 110 Hz. The voltage and current phasors are recorded and the voltage is lowered to zero. This procedure is repeated for all frequencies. For the cable segments with fault at 100 %, the CPC 100 is used to determine the positive and zero sequence impedances. After a complete

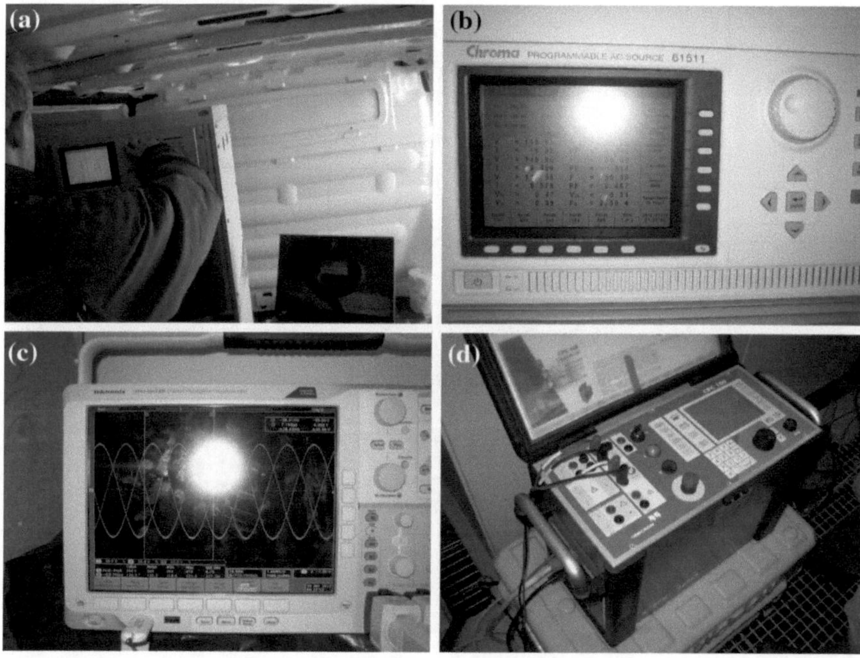

Fig. 5.11 **a** The van with the Chroma 61511 and **b** Displayed results from the Chroma 61511, **c** Oscilloscopes used to monitor waveforms and **d** Impedance measurements by use of the Omicron PCP100

measuring cycle, a new fault type is applied at the fault location and the measuring run is repeated. This is done until all fault impedances have been measured. The person in the field moves to a new fault location and all the above-mentioned steps are repeated. In Fig. 5.11, pictures from the measurements are shown.

5.4 Simulation Model Setup

The Anholt cable system is implemented in DIgSILENT using the cable system interface (ElmCabSys). As mentioned, the cable's electric parameters and physical dimensions are given as input to the model. In Fig. 5.12, the implementation of the first major section of the cable system is shown.

From the figure, it can be seen that six line segments per major section are used for the core, sheath and ECC. This is necessary as the ECC is shifted in the middle of each minor section.

Three single-phase sources are connected between core and sheath and to a 0.1 Ω resistance connected to neutral ground. The use of a two-terminal source ensures a direct path between the sheaths and the source as in real life.

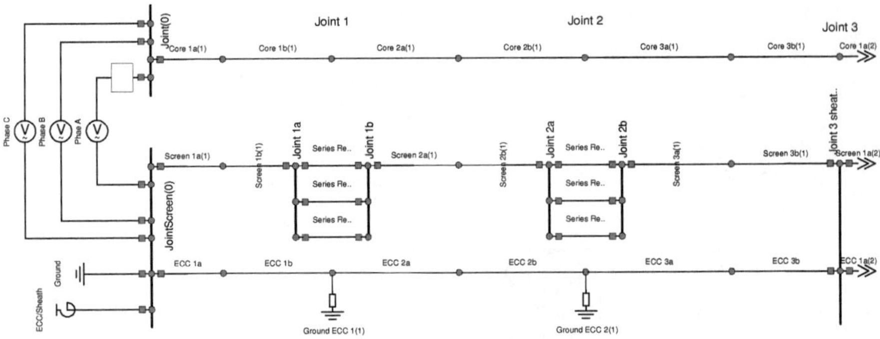

Fig. 5.12 DIgSILENT implementation of the Anholt cable system

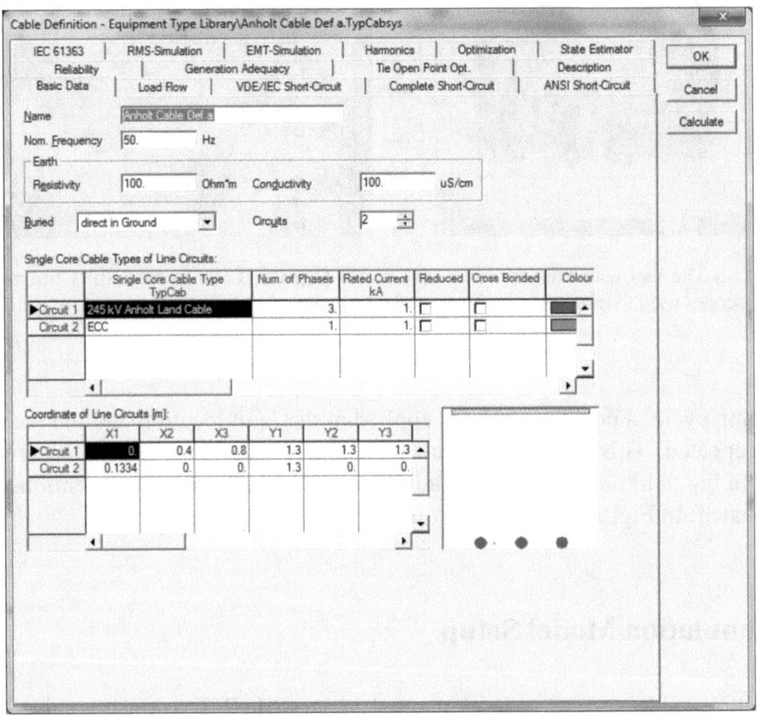

Fig. 5.13 DIgSILENT cable system definition of the Anholt land cable

The configuration of the cables and the ECC is defined in the cable definition system. The input to the definition is seen in Fig. 5.13.

Two cable system definitions must be used because the ECC is shifted in the middle of each minor section. For the second cable system definition, the ECC is placed at 0.6667 m. In Fig. 5.14, the cable definition of the single-core Anholt cable is shown. Another cable definition is defined for the ECC (not shown here).

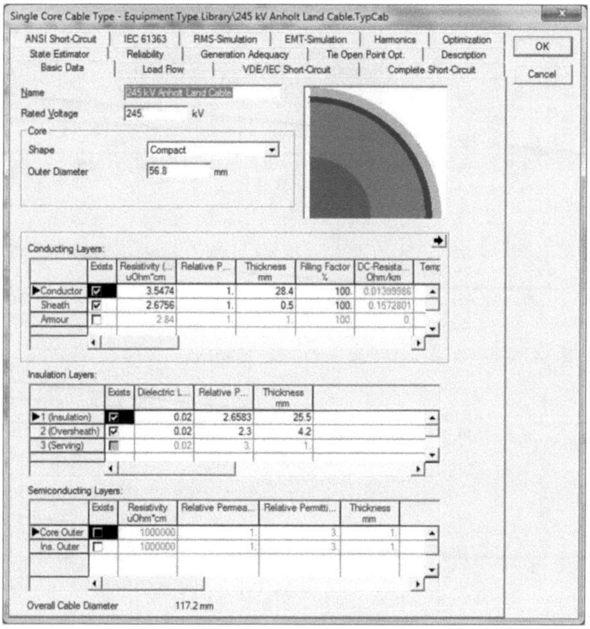

Fig. 5.14 DIgSILENT cable definition of the Anholt land cable

The fault loop impedance of all fault types described in Sect. 5.3 are calculated using the DIgSILENT model. The results are initially used to predict settings for the Chroma power source as the phase voltage has to be set manually. It is of interest to apply the maximum voltage possible to reduce the influence of noise. The apparent power is limited to 4 kVA per phase. In Fig. 5.15, as an example, the current, active-, reactive- and apparent power are plotted as a function of phase voltage for a two phase to sheath to ground fault at Joint 27 seen from Joint 33. The green line in Fig. 5.15d represents the maximum apparent power.

Figure 5.15 shows that the apparent power of Phase A sets the limit for this particular short circuit. At 142 V, the apparent power reaches 4 kVA and no further voltage increase is possible. Such calculations are performed for each short circuit at each frequency.

5.5 Results

The measuring equipment was set up at station Trige (Joint 0) and the measurements involved in Case Study 1 were conducted first (see Fig. 5.2). As an initial test, the positive and zero-sequence impedances were measured using the Chroma power source. Due to the unbalanced system, the impedance of each phase is measured

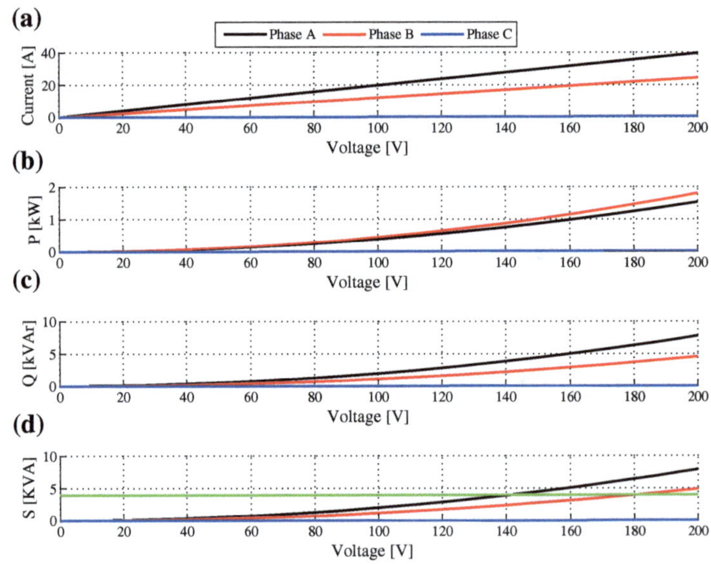

Fig. 5.15 **a** current, **b** active power, **c** reactive power and **d** apparent power as fundtion of voltage in the case of a two phase to sheath to ground fault at Joint 27 seen from Joint 33

Table 5.3 Calculated and measured positive- and zero-sequence impedances for the cable sections from Joint 0 to Joint 27

Impedance [Ω]	Calculated [Ω]	Measured [Ω]
Z_{1A}	6.48 ∠ 75.91°	5.95 ∠ 74.8°
Z_{1B}	5.30 ∠ 83.54°	5.94 ∠ 83.05°
Z_{1C}	6.30 ∠ 93.2°	7.23 ∠ 137.7°
Z_1	6.03 ∠ 84.22°	6.37 ∠ 98.52°
Z_{0A}	5.57 ∠ 28.31°	4.84 ∠ 34.2°
Z_{0B}	5.69 ∠ 26.38°	5.62 ∠ 26.9°
Z_{0C}	5.57 ∠ 28.31°	5.62 ∠ 27.5°
Z_0	5.61 ∠ 27.66°	5.36 ∠ 22.24°

and the positive and zero-sequence impedances are found as the arithmetic mean of the three phase impedances. The measured and simulated impedances are shown in Table 5.3 and the core and sheath currents are shown on phasor form in Fig. 5.16.

Due to crossbonding and the flat formation used, it is expected that the middle phase will have a smaller impedance compared to the impedance of the two outer lying cables and that the magnitude of the two outer lying cables will be approximately equal when a positive sequence voltage is applied. Furthermore, it is expected that the impedance per phase when the system is applied a zero sequence voltage will have an almost equal magnitude and angle. This result is obtained using the model implemented in DIgSILENT, but not when performing the measurements.

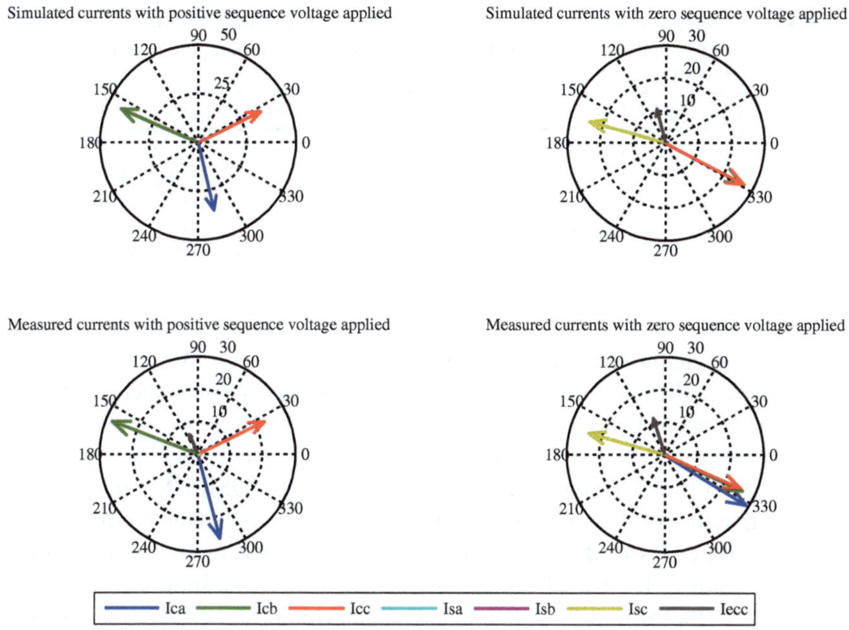

Fig. 5.16 Calculated and measured phasors when applying a positive and zero sequence voltage on the Anholt system and when simulating the impedances in DIgSILENT

Table 5.3 and Fig. 5.16 show that when a positive sequence voltage is applied, the measured current of phase C is smaller; thus, the phase impedance is larger compared to the two additional phases (Z_{1C} underlined in Table 5.3). The phase impedance of phase A is smaller than the two additional phases when a zero-sequence voltage is applied the three-phases (Z_{0A} underlined in Table 5.3).

This behaviour is not expected from a healthy cable system and based on the measuring results, further simulations and an analysis conducted using DIgSILENT, it was concluded that the behaviour could be caused by the ECC, if this conductor was not shifted along the cable route as planned.

If the ECC is laid with phase A the entire cable route, the impedance measured of that particular phase is reduced when a zero sequence voltage is applied. Vectorially, the magnetic field due to the ECC current adds to the magnetic field of the phase A sheath current hence reducing the magnetic circulation of cable A. This effectively lowers the zero-sequence impendence for phase A. Phase B and C are less affected as they placed further from the ECC.

When a positive sequence voltage is applied, the impedance of Phase C is higher than for Phase A and B that are approximately equal. The magnetic field created by the current flowing in the ECC will reduce the apparent impedance of Phase A and in some extend in Phase B. As the ECC is moved closer towards cable A, the impedance of that phase and of phase B starts the become unequal with the phase A impedance being the lowest. If the conductor is placed 12 cm from phase A, the

Table 5.4 Calculated and measured impedance per phase for a positive voltage applied with the Chroma and for the impedances obtained using the seven-step method performed with both the Chroma and CPC 100

Impedance [Ω]	DigSILENT	Chroma	Seven-step with CPC 100	Seven-step with Chroma
Z_{1A}	6.50 ∠ 76.0°	7.14 ∠ 78.28°	7.69 ∠ 84.3°	7.75 ∠ 85.9°
Z_{1B}	5.28 ∠ 83.7°	5.9 ∠ 85.06°	4.97 ∠ 83.9°	4.99 ∠ 84.5°
Z_{1C}	6.30 ∠ 91.6°	7.03 ∠ 91.83°	7.72 ∠ 85.5°	7.71 ∠ 84.8°
Z_1	6.07 ∠ 83.53°	6.70 ∠ 85.06°	6.8 ∠ 84.6°	6.81 ∠ 85.2°

DIgSILENT model predicts that the impedances of phase A and B are more or less equal as found in the measurements; this indicates an average placement of the ECC.

After the hypothesis regarding the wrongly place ECC was set up, the contractor was contacted and it was confirmed that the ECC had in fact been placed with phase A from Joint 0 to 33. From studies conducted in the DIgSILENT model, it was concluded that the wrongly placed ECC would lead to an unbalancing of the system and an increase in losses due to circulating currents. It was proposed to disconnect the ECC at Joint 6, 12, 18, 24 and 33 which will reduce the influence of the ECC significantly as the current flow in the concoctor is reduced. Because only the cable sections from Joint 0 to 33 were installed at the moment of the measuring run, the cable sections from Joint 33 to 51 could be laid without the ECC.

5.5.1 Case Study 1

After the ECC had been disconnected at the joints proposed, the positive sequence impedance was measured again. To ensure correct results, the measurements were done with the Chroma applying a positive sequence voltage, with the CPC 100 using the seven-step method described in Appendix C and as an additional measurement set, the Chroma was used to perform the seven-step impedance measurements as well. The results from all measurements together with the results obtained using DIgSILENT on the repaired Anholt system are presented in Table 5.4.

The measurements performed using the Chroma show that the impedances per phase when applying a positive and zero-sequence voltage are now behaving as expected and the method of reducing the influence of the ECC by disconnecting it at several joints is verified. Furthermore, as a side note, it can be seen that the impedances measured by the Chroma are quite different from the results obtained using the seven-step method with both the CPC 100 and the Chroma. Using the Chroma, angles between 78.28° and 91.83° are measured where the angles determined using the seven-step method are equal between all three phases. Unequal angles are calculated using the DIgSILENT model as well. The error is caused by the assumption of a perfectly balanced system on a system laid in flat formation.

It is noticed from Table 5.4 that the mean values of all measured impedances (Z_1) are almost equal (underlined values in Table 5.4). This indicates that the

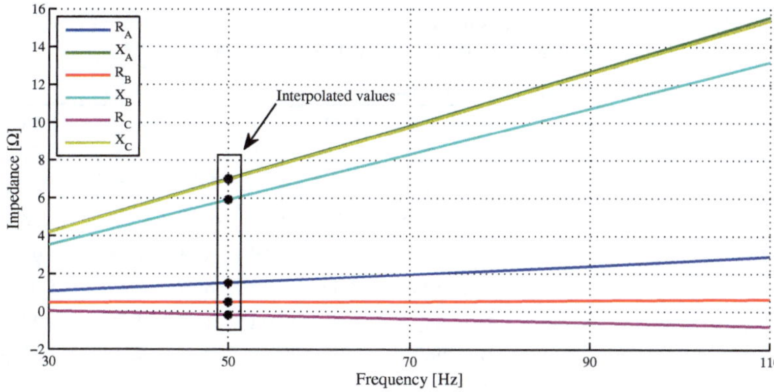

Fig. 5.17 Determination of 50 Hz impedances based upon interpolated measurements conducted at 30, 70, 90 and 110 Hz

seven-step method gives fair results when determining the positive and zero sequence impedances on unbalanced systems as well as on balanced system.

The Chroma was used to measure all fault impedances at 30, 70, 90 and 110 Hz. As an example, the impedances measured for a three phase short circuit are shown in Fig. 5.17.

A small deviation in the linear relationship for the resistive component is noticed, and all impedances are interpolated using only the 30 and 70 Hz impedances. The same procedure is proposed by Omicron and used by the CPC 100 as well. Measurements of all additional fault types for Case Study 1 are presented in Table 5.5.

As already indicated by the initial measurements on the faulted Anholt system, the DIgSILENT model predicts the general behaviour of the crossbonded cable system well, but the accuracy of the fault loop impedances is not high in all cases. A total of 92 short circuits are applied the Anholt cable system for the six cable studies and the results of the additional measurements are presented in Appendix A. The absolute deviation and deviation in percentage of the real and imaginary parts of the fault loop impedances are presented in histograms in Figs. 5.18 and 5.19.

In general; the DIgSILENT model predicts the behaviour of the Anholt system well, but the accuracy of the individual fault loop impedances varies significantly. Neural networks can therefore not be trained using the data and can therefore not be used for fault location on the system. The deviation in the real part of the fault impedance is between −200 to 133 % where the deviations in the reactive part are smaller, but still significant (−61 to 38 %). The large relative errors are in general obtained when either the real or imaginary part is small as it for instance is the case of the real part of Z_{1C} in Case Study 2 presented in Table A.1. Here, the measured real part is −0.03 Ω and the calculated is −0.09 Ω. The absolute deviation is small, but the deviation in percentage becomes 200 %.

Table 5.5 Calculated fault impedances for the cable sections from Joint 0 to Joint 27 with the fault location at Joint 27 after disconnection of the ECC at Joint 6, 12, 18, 24 and 33

Fault type	Impedance [Ω]	Calculated [Ω]		Measured [Ω]	
Phase ABC/S (positive-seq)	Z_{1A}	6.50 ∠ 76.0°	1.53 + j6.31	7.14 ∠ 78.28°	1.45 + j7.00
	Z_{1B}	5.28 ∠ 83.7°	0.68 + j5.31	5.9 ∠ 85.06°	0.51 + j5.90
	Z_{1C}	6.30 ∠ 91.6°	−0.17 + j6.36	7.03 ∠ 91.83°	−0.23 + j7.03
	Z_1	6.07 ∠ 83.53°	0.68 + j5.99	6.70 ∠ 85.06°	0.58 + j6.64
Phase AB/S	$Z_{f,A}$	4.72 ∠ 74.7°	1.25 + j4.55	5.06 ∠ 78.8°	0.98 + j4.97
	$Z_{f,B}$	7.92 ∠ 70.2°	2.69 + j7.45	8.82 ∠ 68.3°	3.03 + j7.61
Phase BC/S	$Z_{f,B}$	4.21 ∠ 69.2°	1.50 + j3.94	4.57 ∠ 74.5°	1.22 + j4.40
	$Z_{f,C}$	9.6 ∠ 79.3°	1.78 + j9.43	9.92 ∠ 73.6°	2.81 + j9.51
Phase CA/S	$Z_{f,A}$	11.37 ∠ 61.0°	5.52 + j9.94	11.26 ∠ 56.0°	6.29 + j9.33
	$Z_{f,C}$	4.90 ∠ 74.2°	1.33 + j4.71	5.09 ∠ 78.9°	0.98 + j4.99
Phase A/S	$Z_{f,A}$	5.65 ∠ 66.7 °	2.23 + j5.19	5.94 ∠ 70.4°	1.99 + j5.59
Phase B/S	$Z_{f,B}$	4.78 ∠ 61.2°	2.30 + j4.19	5.12 ∠ 66.5°	2.04 + j4.69
Phase C/S	$Z_{f,C}$	5.61 ∠ 65.5°	2.33 + j5.10	5.85 ∠ 69.6°	2.04 + j5.48
Zero-sequence	Z_{0A}	6.04 ∠ 23.2°	5.55 + j2.38	5.44 ∠ 24.9°	4.93 + j2.29
	Z_{0B}	5.94 ∠ 21.6°	5.52 + j2.19	5.48 ∠ 24.2°	5.00 + j2.25
	Z_{0C}	5.95 ∠ 24.1°	5.43 + j2.43	5.33 ∠ 25.0°	4.83 + j2.26
	Z_0	5.98 ∠ 23.0°	5.50 + j2.33	5.42 ∠ 24.7°	4.92 + j2.26

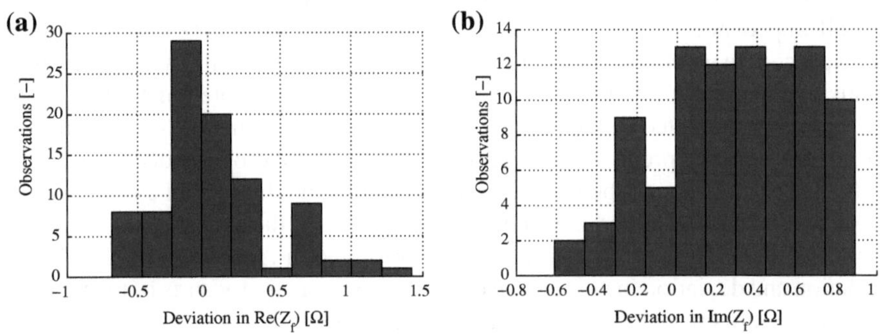

Fig. 5.18 Absolute deviations in the **a** real and **b** imaginary parts of the fault loop impedance for measurements conducted on the Anholt land cable

5.5.2 Discussion

The deviations are most likely found in a simplified representation of the zero-sequence system. Zero-sequence current flows in case of all applied short circuits wherefore it characteristics becomes of importance. Including a very detailed representation of the sheath system using state-of-the-art models is not sufficient to obtain accuracy results. The laying of the cable is also known to cause deviations as it often is not as ideal as when modelled. The lengths of the cable sections can vary and small

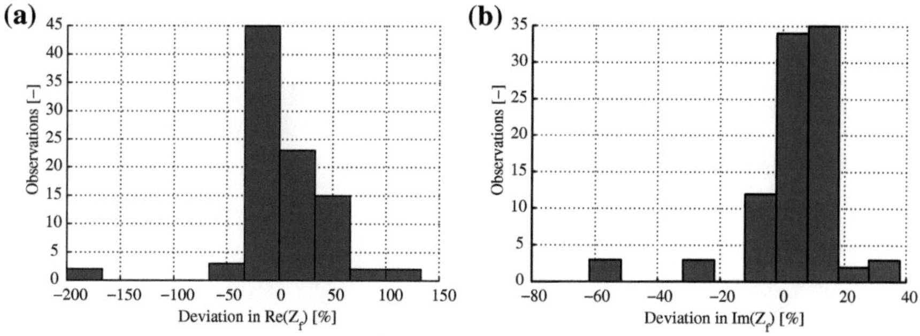

Fig. 5.19 Deviation in percentage in the **a** real and **b** imaginary parts of the fault loop impedance for measurements conducted on the Anholt land cable

errors in the cable physical description will cause errors when the impedance and admittance matrix are formed. These errors may be small, but the accuracy required for fault location studies is very high, and they will thus have an effect as well.

If has been well know that impedance calculations on especially cable systems and especially the zero-sequence system is difficult in practice. The findings presented in this chapter indicate that this is still the case even with the most advanced models available today. The results presented are, however, only valid for the Anholt system and further studies of the same type on one or two more systems is of interest to confirm the results. It is seems, however, not likely that an accuracy useable for fault location studies can be achieve in any case and for this thesis, the impedance-based method is rejected.

5.5.3 Conclusions on the Impedance-Based Field Measurements

From the results of the six case studies conducted it can be concluded that the DIgSILENT model predicts the behaviour of the Anholt cable system under fault conditions well, but the accuracy on the fault loop impedance is low. The large errors indicate that adjustments of the model's parameters cannot solve the problem alone so an accuracy applicable for fault location can be accomplished. Therefore, in conclusion:

> *A neural network cannot be trained to estimate the fault location on crossbonded cables because accurate training data cannot be provided by state-of-the-art simulation models.*

Reference

1. Cigré Task Force B1.26, *Earth Potential Rises in Specially Bonded Screen Systems* (Paris, France, Cigré, 2008)

Part III
Fault Location on Crossbonded Cables Using Travelling Waves

Chapter 6
Wave Propagation on Three Single-Core Solid-Bonded and Crossbonded Cable Systems

In case of an internal ideal cable fault, the voltage will instantaneously drop in the faulted phase at the fault location. This is equivalent to applying a voltage impulse equal, but with opposite sign to that existing on the faulted phase at the time of fault. The study of how high frequency impulses propagate on a crossbonded system is therefore interesting for fault location purposes and is studied in this chapter.

As previously explained, for longer cables at high voltage levels, it is normal procedure to use a sheath crossbonding scheme. Due to the crossbonding, discontinuities in the surge impedance occur along the cable route. These discontinuities cause reflections and thereby complicate fault location, especially when using the single ended method [1].

In order to reduce the complexity of the analysis carried out, this chapter is split into two main sections. In the first section, a study of wave propagation on a cable system with single core unarmoured cables is carried out (no crossbonding). The modal theory is introduced and used to explain the system's behaviour. In the second section, wave propagation on the same cable system, but with the sheath crossbonded is presented using the knowledge gained in section one.

6.1 Wave Propagation on Three Single Core Solid-Bonded Cable System

Because the conductors of a three single core cable are electromagnetically coupled, a single propagation constant cannot be calculated for each conductor. As a result, a single wave velocity or attenuation constant does not exist either. As shown in Eqs. 2.1 and 2.2, a unique wave velocity is needed for the two classical online fault location methods based on travelling waves. To simplify the analysis of transmission lines, the modal decomposition theory can, however, be used [2–5]. This method utilises frequency dependent eigenvalues and transformation matrices allowing the n conductor transmission system to be represented by n independent modes [6].

C. F. Jensen, *Online Location of Faults on AC Cables in Underground Transmission Systems*, Springer Theses, DOI: 10.1007/978-3-319-05398-1_6,

The characteristics of these modes must be examined before a suitable fault location method can be developed for crossbonded cables.

As a starting point, the series impedance and shunt admittance matrices are defined. Equation 6.3 shows how the series impedance matrix is defined in relation to the core and sheath voltage and currents for a three single core unarmoured cable.

$$I_{C,t} = \frac{U_{C,t}}{Z_{0C}} \tag{6.1}$$

$$I_{S,t} = -I_{C,t} \tag{6.2}$$

$$\begin{bmatrix} U_{C_1} \\ U_{S_1} \\ U_{C_B} \\ U_{S_B} \\ U_{C_C} \\ U_{S_C} \end{bmatrix} = \begin{bmatrix} Z_{C_AC_A} & Z_{C_AS_A} & Z_{C_AC_B} & Z_{C_AS_B} & Z_{C_AC_C} & Z_{C_AS_C} \\ Z_{S_AC_A} & Z_{S_AS_A} & Z_{S_AC_B} & Z_{S_AS_B} & Z_{S_AC_C} & Z_{S_AS_C} \\ Z_{C_BC_A} & Z_{C_BS_A} & Z_{C_BC_B} & Z_{C_BS_B} & Z_{C_BC_C} & Z_{C_BS_C} \\ Z_{S_BC_A} & Z_{S_BS_A} & Z_{S_BC_B} & Z_{S_BS_B} & Z_{S_BC_C} & Z_{S_BS_C} \\ Z_{C_CC_A} & Z_{C_CS_A} & Z_{C_CC_B} & Z_{C_CS_B} & Z_{C_CC_C} & Z_{C_CS_C} \\ Z_{S_CC_A} & Z_{S_CS_A} & Z_{S_CC_B} & Z_{S_CS_B} & Z_{S_CC_C} & Z_{S_CS_C} \end{bmatrix} \begin{bmatrix} I_{C_A} \\ I_{S_A} \\ I_{C_B} \\ I_{S_B} \\ I_{C_C} \\ I_{S_C} \end{bmatrix} \tag{6.3}$$

where Z_{CiCi} is the self impedance of core of cable i, Z_{SiSi} is the self impedance of the sheath of cable i, and Z_{CiSm} is the mutual impedance between core of cable i and the sheath of cable m.

The shunt admittance matrix for a six conductor system is defined as [7]:

$$\begin{bmatrix} I_{C_A} \\ I_{S_A} \\ I_{C_B} \\ I_{S_B} \\ I_{C_C} \\ I_{S_C} \end{bmatrix} = \begin{bmatrix} Y_{C_AC_A} & Y_{C_AS_A} & 0 & 0 & 0 & 0 \\ Y_{S_AC_A} & Y_{S_AS_A} & 0 & 0 & 0 & 0 \\ 0 & 0 & Y_{C_BC_B} & Y_{C_BS_B} & 0 & 0 \\ 0 & 0 & Y_{S_BC_B} & Y_{S_BS_B} & 0 & 0 \\ 0 & 0 & 0 & 0 & Y_{C_CC_C} & Y_{C_CS_C} \\ 0 & 0 & 0 & 0 & Y_{S_CC_C} & Y_{S_CS_C} \end{bmatrix} \begin{bmatrix} U_{C_A} \\ U_{S_A} \\ U_{C_B} \\ U_{S_B} \\ U_{C_C} \\ U_{S_C} \end{bmatrix} \tag{6.4}$$

where Y_{CiCi} is the self-shunt admittance of core i, Y_{SiSi} is the self-shunt admittance of sheath i, and $Y_{C_iS_m}$ is the mutual admittance between core i and sheath m. The method presented by Ametani in 1980 for calculating impedances and admittance for coaxial cables is used in this thesis [7].

Equations 6.3 and 6.4 show the coupling between the conductors. Using eigenvalue theory, the system can be decoupled thus resulting in a much simpler system to analyse.

6.1.1 Modal Decomposition

The modal transformation matrices are frequency dependent and complex. This is due to the frequency dependence nature of the shunt admittance and series impedance matrices. The frequency dependence of the transformation matrices is often neglected for OHL-systems because of the limited variation over the frequency range of interest and the almost negligible imaginary part of the matrix components [8]. Neither are, however, true for cable systems—only at very high frequencies, these assumptions hold true.

The eigenvalues of the cable system are calculated with modal transformation theory according to Eq. 6.5 [3].

$$\mathbf{S}^{-1}\mathbf{ZYS} = \lambda \tag{6.5}$$

where λ is the systems eigenvalues on diagonal form, **S** is formed by the eigenvectors associated with the eigenvalues of the matrix **ZY** in a diagonal form and **Z** and **Y** are the series impedances and shunt admittance matrices of the cable system. The modal currents $\underline{I}_m$ are obtained as shown in Eq. 6.8.

$$\underline{I}_m = \mathbf{Q}^{-1}\underline{I}_p \tag{6.6}$$

$$\underline{U}_m = \mathbf{S}^{-1}\underline{U}_p \tag{6.7}$$

$$\mathbf{Q}^{-1} = \mathbf{S}^T \tag{6.8}$$

where $\underline{I}_p$ are the phase domain currents defined as $\underline{I}_p = [I_{cA}, I_{sA}, I_{cB}, I_{sB}, I_{cC}, I_{sC}]^T$ with I_{ci} and I_{si} being the core and sheath currents of cable i, and $\mathbf{Q}^{-1}$ is equal to $\mathbf{S}^T$ and $\mathbf{S}^T$ denotes the matrix transpose of **S**. The phase currents are found by rearranging Eq. 6.8 as shown in 6.9.

$$\underline{I}_p = \mathbf{Q}\underline{I}_m \tag{6.9}$$

By substituting the modal and phase domain currents vectors $\underline{I}_m$ and $\underline{I}_p$ by the modal and phase domain voltages $\underline{U}_m$ and $\underline{U}_p$ and using the transformation matrix **S** instead of **Q**, the voltage transformation is obtained.

The propagation constant $\underline{\gamma}_k$ is calculated as the square root of the kth eigenvalue. The propagation constant is a complex number with the following definition:

$$\underline{\gamma}_k = \sqrt{\underline{\lambda}_k} = \alpha_k + j\beta_k \tag{6.10}$$

where α_k is the attenuation constant and β_k is the phase coefficient. The wave velocity is calculated as:

$$v_k = \frac{\omega}{\beta_k} \tag{6.11}$$

A phenomenon called 'Eigenvalue Switching' can occur if a standard QR-method is used to solve Eq. 6.5. The eigenvalue calculated at one frequency does not always corresponds to the eigenvalue calculated at the next frequency. The eigenvalue versus frequency plot is smooth, but do to the problems associated with the QR-based solution, this is not always the case. The eigenvalue switching problem is solved by adopting a Newton Raphson (NR) solution method as proposed in [9]. This method ensures a smooth eigenvalue versus frequency behaviour. The NR method is used in this work to perform any modal calculation.

6.1.2 Modal Wave Propagation Characteristics

Two typical laying configurations are used in Denmark. The cables are either laid in touching trefoil or in flat formation. How the different laying arrangements affect the cable systems' propagation characteristics is interesting for fault location purposes and is therefore examined in this section. It is of special interest to examine which cable and cable system parameters affect the wave propagation velocities. It will be very beneficial to choose a propagation mode with a frequency independent velocity, one that does not depend on parameters that change as function of time, changed along the cable route or that is difficult to determine in practice.

6.1.3 Trefoil Formation

Firstly, the modal decomposition of the case study three single core 165 kV cable laid in touching trefoil formation is examined. The cable and laying configurations were described in Sect. 4.1.

The frequency dependent transformation matrix **Q** is calculated for the system at frequencies from 1 Hz to 1 MHz by using Eq. 6.5. At high frequencies, the modes become decoupled, and evaluating **Q** at 1 MHz and taking the real part (neglectable imaginary parts) yields the result shown in Eq. 6.12.

$$\mathbf{Q} = \begin{bmatrix} 0 & 1.00 & 1.00 & -0.50 & 0 & 0 \\ 1.00 & -1.00 & -1.00 & 0.50 & 1.00 & -0.50 \\ 0 & 1.00 & 0 & 1.00 & 0 & 0 \\ 1.00 & -1.00 & 0 & -1.00 & 0 & 1.00 \\ 0 & 1.00 & -1.00 & -0.50 & 0 & 0 \\ 1.00 & -1.00 & 1.00 & 0.50 & -1.00 & -0.50 \end{bmatrix} \tag{6.12}$$

Mode 1, represented by column 1, is a classical ground mode with the current returning in the ground. Columns 2–4 in **Q** show three coaxial modes with all current propagating between the core and metal sheath. Columns 5–6 are intersheath modes. A visual interpretation of the modes is presented in Fig. 6.1.

Calculating the attenuation and velocity based on Eq. 6.5, the definition given in Eqs. 6.10 and 6.11 gives the result shown in Fig. 6.2.

The frequency dependence of the coaxial modes is clearly seen from 1 Hz to approximately 10 kHz. At 1 kHz, these modes become equal, and from above 10 kHz, the velocities can be assumed constant because the metallic sheath provides complete shielding, and thus no field exists outside this layer. At low frequencies, complete coaxial modes do not exist due to interaction between other conductors and earth.

The two intersheath modal velocities are equal due to the equal size of current loops they represent (see Fig. 6.1). The velocity of the intersheath modes are frequency independent above 100 Hz. The attenuation of the intersheath mode waves is

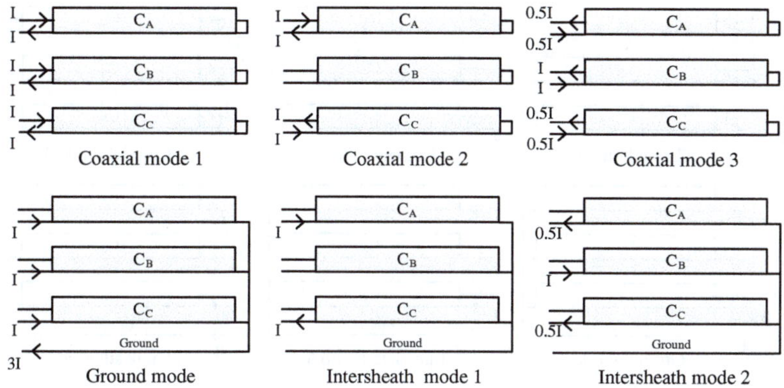

Fig. 6.1 Visual presentation of the three coaxial mode currents used for the modal decomposition of three single core cables laid in touching trefoil

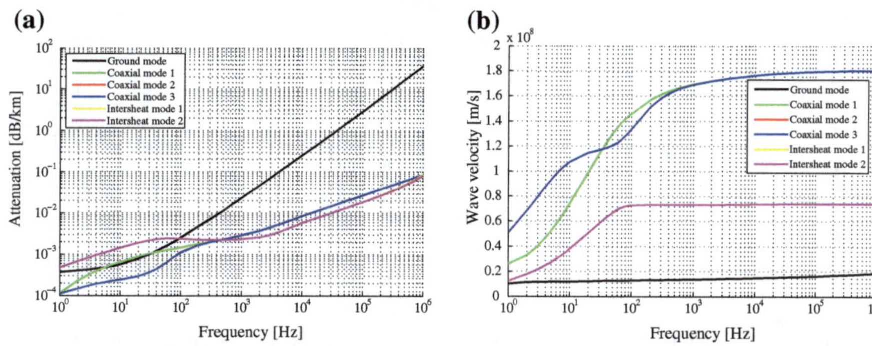

Fig. 6.2 **a** Modal attenuation and **b** velocity as function of frequency for three single core cable laid in touching trefoil

higher than for the coaxial modes at high frequencies. Attenuation of high frequency components can be a problem as the lack of these frequency components makes the exact arrival instance of the fault waves harder to determine.

6.1.4 Flat Formation

The modal decomposition for the case study cable system laid in flat formation is examined next (see Sect. 4.1). The transformation matrix **Q** evaluated at 1 MHz is shown in Eq. 6.13.

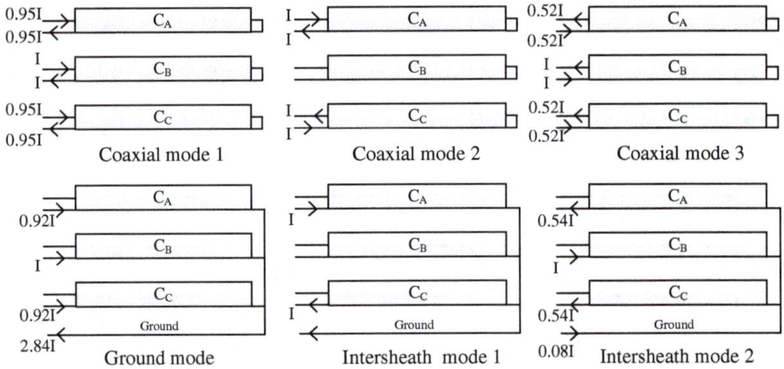

Fig. 6.3 Visual presentation of the three coaxial mode currents used for the modal decomposition of three single core cables laid in flat formation

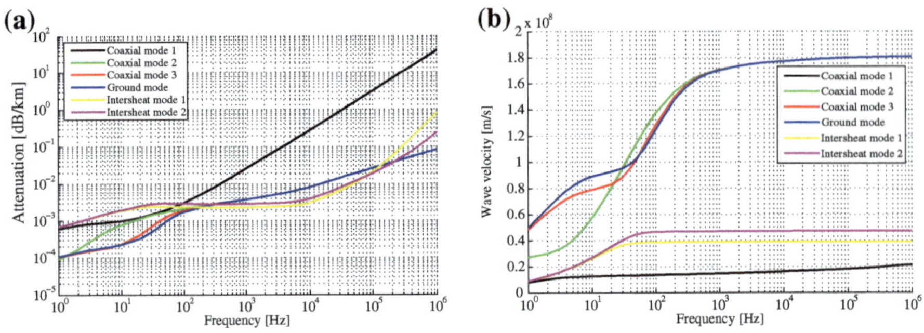

Fig. 6.4 **a** Modal attenuation and **b** velocity as a function of frequency for three single core cables laid in flat formation

$$\mathbf{Q} = \begin{bmatrix} 0.95 & 1.00 & -0.52 & 0 & 0 & 0 \\ -0.95 & -1.00 & 0.52 & 0.92 & 1.00 & -0.54 \\ 1.00 & 0 & 1.00 & 0 & 0 & 0 \\ -1.00 & 0 & -1.00 & 1.00 & 0 & 1.00 \\ 0.95 & -1.00 & -0.52 & 0 & 0 & 0 \\ -0.95 & 1.00 & 0.52 & 0.92 & -1.00 & -0.54 \end{bmatrix} \tag{6.13}$$

It is noticed that column 1–3 are now coaxial modes, column 4 is the ground mode, and columns 5 and 6 are intersheath modes. A visual interpretation of the modes are presented in Fig. 6.3.

The modes bear close resemblance to the ones determined for a cable system laid in touching-trefoil formation with three perfect coaxial modes, two intersheath and one ground mode. Due to unsymmetry, the ground is excited for Intersheath mode 2 contrary to what was the case for the system in touching-trefoil formation. The modal velocities and attenuation is computed from 1 Hz to 1 MHz and the result is shown in Fig. 6.4.

Table 6.1 Modal velocities for a flat and touching trefoil formation evaluated at 1 MHz

Mode	Flat formation [m/s]	Touching trefoil [m/s]
Coaxial mode 1–3	1.8118×10^8	1.8119×10^8
Ground mode	0.1995×10^8	0.1541×10^8
Intersheath mode 1	0.3946×10^8	0.6155×10^8
Intersheath mode 2	0.4727×10^8	0.6155×10^8

Table 6.2 Travel time across one 2 km minor section for a coaxial, an Intersheat 1 and 2, and a ground mode wave

Modal wave type	Variable	Travelling time [μs]
Coaxial wave group	τ_{CG}	11.03
Intersheath 1	τ_{I1}	50.68
Intersheath 2	τ_{I2}	42.1
Ground mode	τ_G	106.38

For unsymmetrically laid single core cables, the number of different modal velocities in the system at high frequencies are increased by one compared to a symmetrically laid system. This is an important result if one chooses to use the intersheath modal velocities for fault location purposes. The asymptotically modal velocities for the flat and touching trefoil formation evaluated at 1 MHz are summarised in Table 6.1.

After the modal decomposition, one of the modal velocities shown in Table 6.1 can be used as input in Eq. 2.1 or 2.2. An advantage already seen for using the coaxial mode velocity is that the coaxial modal velocity is independent of the placement of the cables—the intersheath and ground mode wave velocities are not.

6.1.5 Pulse Propagation on a Three Single-Core Solid-Bonded Cable System

As mentioned in the introduction to this chapter, a voltage impulse applied between core(s) and sheath(s) of a cable system will result in the same wave propagation characteristics as fault created waves do. Because an impulse test can be carried out on a none-energised system, the analysis becomes much simpler as no power frequency signal is present.

The same 165 kV case study cable system laid in flat formation and described in Sect. 4.1 is used. The underground link (UG-link) length is chosen as 4 km, and simulations of the core and sheath voltages and currents are carried out at the sending end, the cable middle and the receiving end (0, 2 and 4 km).

A 10 kV impulse voltage created by using two exponential curves and a surge impedance of 12 Ω is applied. The sheaths are bonded and grounded through 10 Ω resistors at each cable end. The impulse is applied between the core of phase A and the three bonded sheaths.

To obtain measurable currents, the three cores at Terminal B and the two non-energised cores at terminal A are grounded through 500 Ω resistance.

As shown in Fig. 6.4, three coaxial waves will travel in the system. At high frequencies they travel with the same velocity and will because of this arrive at any location at the same time. They will in the following be considered as one wave group and named 'Coaxial Wave Group'. Table 6.2 shows the travelling time across one 2 km cable section (half the cable) for a Coaxial wave group, an Intersheat mode 1 and 2 wave and a ground mode wave.

The core and sheath voltages are presented in Fig. 6.5, where it is seen that a coaxial wave travelling at 1.8118×10^8 m/s will need τ_{CG} (11.03 μs) to reach the cable middle (2 km). At $2\tau_{CG}$, the wave is seen at the cable end (Fig. 6.5e) where it is reflected with a positive reflection coefficient and seen again at the cable middle at $3\tau_{CG}$. At Terminal A, the wave is reflected with a negative coefficient and seen at the cable middle at $5\tau_{CG}$. The coaxial wave continues this way until it is damped out. The core and sheath currents at Terminal A the cable middle, and Terminal B are shown in Fig. 6.6.

The coaxial current wave travels in the same manner as the voltage waves, but the polarity of the reflection coefficients at the cable ends are reversed. On the sheath of the energised phase, a current wave equal, but with opposite sign to the current existing on the conductor is observed. This current is generated by magnetic induction, and the sign is determined so that the resulting magnetic field opposes the original change in magnetic flux (Lenz law). The AC current density J in a conductor decreases exponentially from its surface value J_S according to the depth d from the surface as shown in Eq. 6.14.

$$J = J_s e^{-d/\delta} \tag{6.14}$$

where δ is the skin depth defined as the depth at which the current density is reduced to $1/e$. The skin depth can be approximated as shown in Eq. 6.15.

$$\delta = \sqrt{\frac{2\rho}{\omega\mu}} \quad [\mathrm{m}] \tag{6.15}$$

where ρ is the resistivity of the conductor, ω is the angular frequency and μ is the magnetic permeability of the conductor material. For a skin depth much smaller than the thickness of the sheath, the field outside the sheath layer can be neglected and therefore, no magnetic coupling to the other cables is present. The skin depth for copper and aluminium at 1 MHz is 65 and 84 μm respectively; much smaller than the typical sheath layer thickness.

The modal voltages can be obtained from the phase domain core and sheath voltages using the transformation matrix $\mathbf{S}^{-1}$ where $\mathbf{S} = (\mathbf{Q}^{-1})^T$. Using a transformation matrix evaluated at 1 MHz, the modal voltages are calculated as shown in Eq. 6.16.

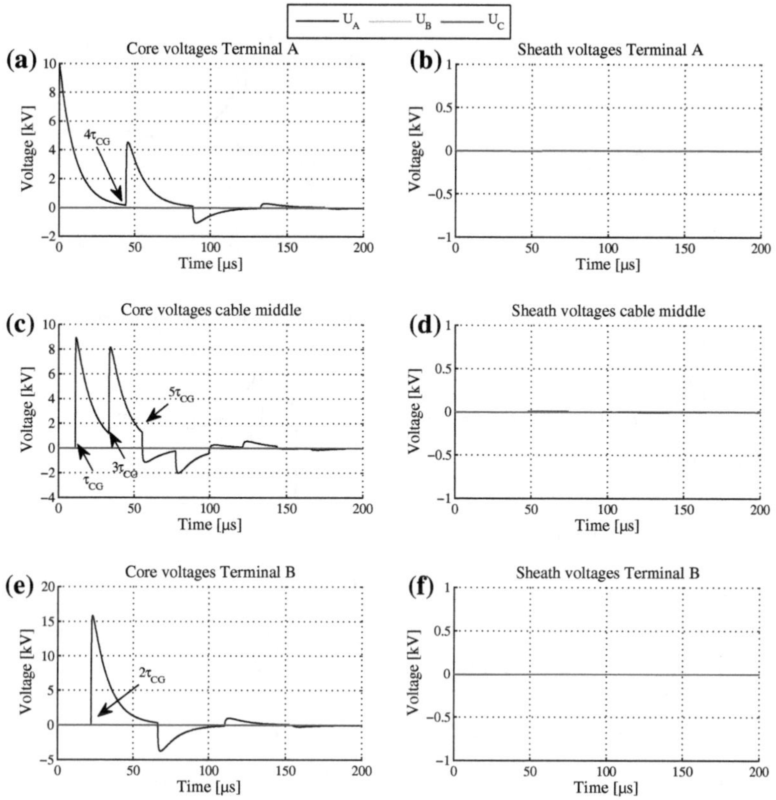

Fig. 6.5 Core and sheath voltage at Terminal A, the cable middle and Terminal B after a 1.2/50 μs voltage impulse is applied phase A

$$\begin{bmatrix} U_{C1} \\ U_{C2} \\ U_{C3} \\ U_G \\ U_{I1} \\ U_{I2} \end{bmatrix} = \begin{bmatrix} 0.9513 & -0.9513 & 1.0000 & -1.0000 & 0.9513 & -0.9513 \\ 1.0000 & -1.0000 & 0 & 0 & -1.0000 & 1.0000 \\ -0.5236 & 0.5236 & 1.0000 & -1.0000 & -0.5236 & 0.5236 \\ 0 & 0.9211 & 0 & 1.0000 & 0 & 0.9211 \\ 0 & 1.0000 & 0 & 0 & 0 & -1.0000 \\ 0 & -0.5409 & 0 & 1.0000 & 0 & -0.5409 \end{bmatrix} \begin{bmatrix} U_{cA} \\ U_{sA} \\ U_{cB} \\ U_{sB} \\ U_{cC} \\ U_{sC} \end{bmatrix} \tag{6.16}$$

where $U_{C1} - U_{C3}$ are the coaxial voltage mode 1–3, U_G is the voltage ground mode and U_{I1} and U_{I2} are voltage intersheath mode 1 and 2. The modal currents are calculated by using $\mathbf{Q}^{-1}$ evaluated at 1 MHz.

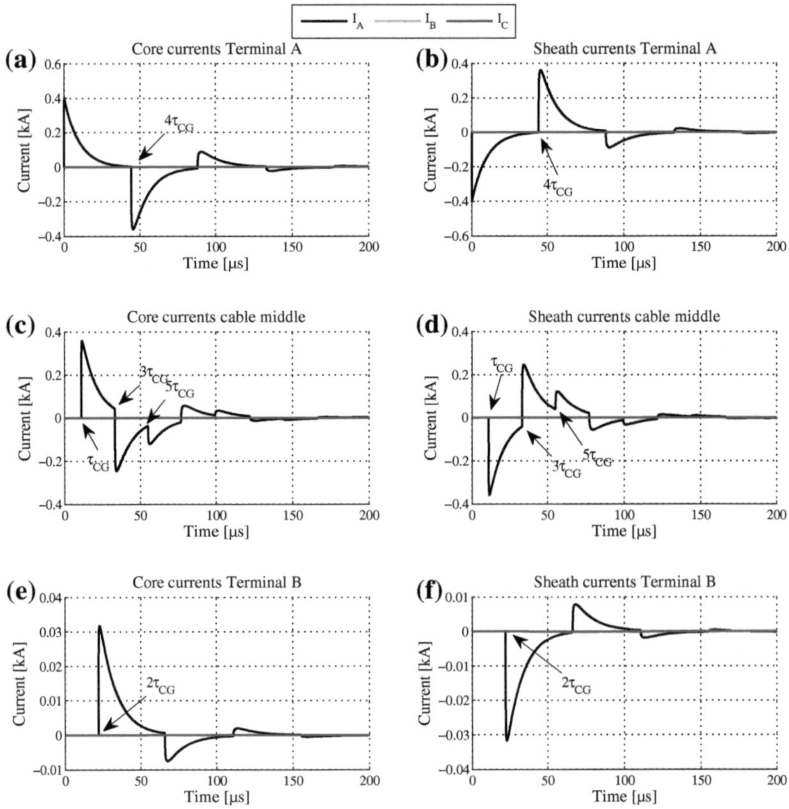

Fig. 6.6 Core and sheath current at Terminal A, the cable middle and Terminal B after a 1.2/50 μs voltage impulse is applied phase A

$$\begin{bmatrix} I_{C1} \\ I_{C2} \\ I_{C3} \\ I_G \\ I_{I1} \\ I_{I2} \end{bmatrix} = \begin{bmatrix} 0.3389 & 0 & 0.3553 & 0 & 0.3389 & 0 \\ 0.5000 & 0 & 0 & 0 & -0.5000 & 0 \\ -0.3389 & 0 & 0.6447 & 0 & -0.3389 & 0 \\ 0.3416 & 0.3416 & 0.3702 & 0.3702 & 0.3416 & 0.3416 \\ 0.5000 & 0.5000 & 0.0000 & -0.0000 & -0.5000 & -0.5000 \\ -0.3416 & -0.3416 & 0.6298 & 0.6298 & -0.3416 & -0.3416 \end{bmatrix} \begin{bmatrix} I_{cA} \\ I_{sA} \\ I_{cB} \\ I_{sB} \\ I_{cC} \\ I_{sC} \end{bmatrix} \tag{6.17}$$

The modal domain voltages and currents at Terminal A, the middle of the cable and at Terminal B are shown in Fig. 6.7.

It is seen from Fig. 6.7 that the modal voltages and current signals consist only of coaxial components. No intersheath or ground mode waves are created at Terminal A when the pulse is applied, and none are therefore seen in the cable middle nor at the cable ends.

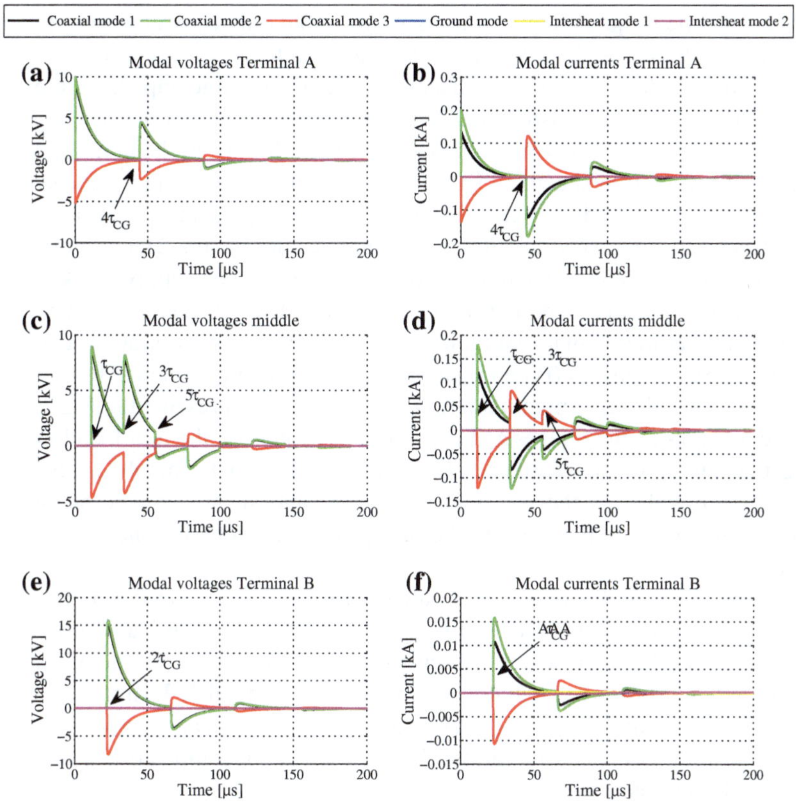

Fig. 6.7 Modal voltages and current at at Terminal A, the middle of the cable and at Terminal B transformed at 1 MHz

The fact that only coaxial waves are created when an impulse is applied between core and sheath is an important finding for fault location purposes. Only coaxial waves will therefore travel directly from the fault location to the fault locator terminals (FLT) where the fault location information can be extracted. The next section examines how waves propagate on crossbonded cables and how the intersheath mode wave contributes to the voltage and current profiles on a crossbonded system.

6.2 Wave Propagation on a Three Single Core Crossbonded Cable

Transient calculations in crossbonded cables have been a study of interest for many years and is considered a complicated field of study [10–14].

It was shown in Sect. 6.1 that a three single core cable will be decomposed into six modes and will therefore have six different modal velocities (the velocities can be equal at some frequencies). It was also shown that only coaxial waves are created when an impulses is applied between core and sheath.

On a crossbonded cable system, discontinuities in the surge impedance exist along the cable route at each crossbonding due to the transposition of the sheaths. It must be examined how this affects the coaxial wave travelling from the fault location and whether and when intersheath and ground mode waves are created in the system.

The 165 kV case study system laid in flat formation presented in Sect. 4.1 is used to construct one major section with crossbonded cables having 2 km minor sections. The same impulse voltage generator used in Sect. 6.1 is used to apply a 10 kV impulse to phase A. The sheaths are bonded and grounded through 10 Ω resistors at each cable end, and in order to obtain measurable currents, the three cores at the far-end and the two non-energised phases (B and C) at the sending end are grounded through 500 Ω resistance. The PSCAD implementation of the system is presented in Fig. 6.8.

All core and sheath voltages and currents are recorded at the four locations shown in Fig. 6.8. These locations are named Terminal 1–4, with Terminal 1 as the sending end and Terminal 4 as the receiving. The core and sheath voltage at these four locations are shown in Fig. 6.9 and the in Fig. 6.10, the core and sheath currents are shown.

The figures show the very complicated nature of wave propagation on a crossbonded cable system. Comparing Figs. 6.9 and 6.10 to Figs. 6.5 and 6.6, it can be seen that many additional waves are travelling on a crossbonded system. The modal voltages and currents are calculated and plotted at Terminal 1, Terminal 2, Terminal 3 and Terminal 4 and the result is shown in Fig. 6.11. Because each of the minor sections have the same configuration, the same modal decomposition matrix can be used for all sections.

Observing Fig. 6.11 one important finding is made: The time is takes a coaxial wave to travel from one location on a crossbonded cable system to another is not affected by the crossbondings if such are encountered. For instance, the first coaxial wave group that arrives at Terminal 4 arrive at $3\tau_{CG}$ (Fig. 6.11g and h).

When a coaxial wave meets a crossbonding, a part of the wave is refracted onto the circuit constituted by the same core but the transposed sheath where it continues as a new coaxal wave. This occurs at every crossbonding wherefore the arrival instance of the wave at any location is the same as if no crossbondings were present. Both the single- and two-terminal methods assumes a waves which has traveled directly from the fault location to the FLT. The coaxial modal waves can therefore be used for fault location on crossbonded cables without any compensation for the effects of the crossbondings.

Seen from Terminal A (sending end), the surge impedances and electrical configurations are equal for the crossbonded cable compared to the solid-bonded cable studies in the last section. The same coaxial waves are therefore sent from Terminal A when an impulse is applied. However, at Terminal 2 and 3, two intersheath mode currents and an intersheath mode 1 voltage wave are seen. As these waves are not created at the sending end when the impulse is applied, they must have been created when the first coaxial wave met the crossbondings at Terminal 2 and 3.

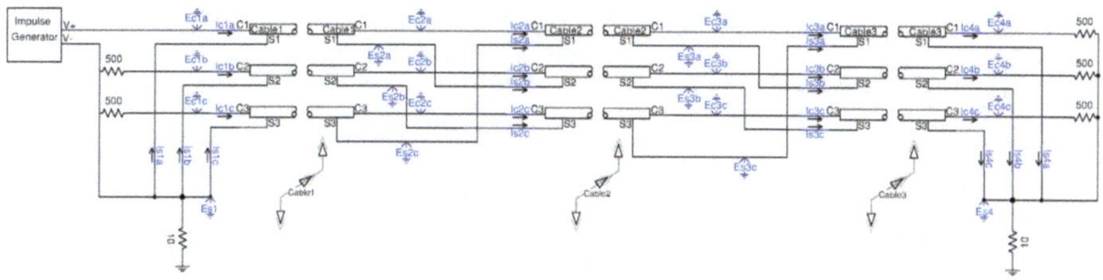

Fig. 6.8 PSCAD implementation of the one major section crossbonded cable system

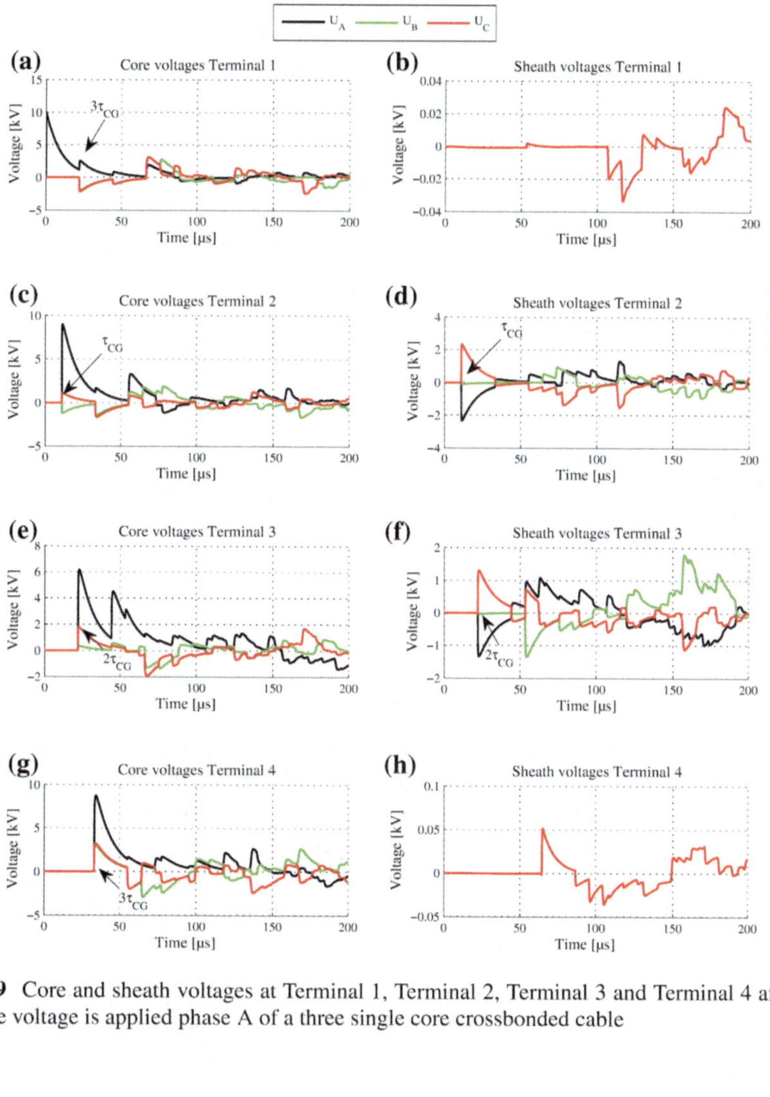

Fig. 6.9 Core and sheath voltages at Terminal 1, Terminal 2, Terminal 3 and Terminal 4 after an impulse voltage is applied phase A of a three single core crossbonded cable

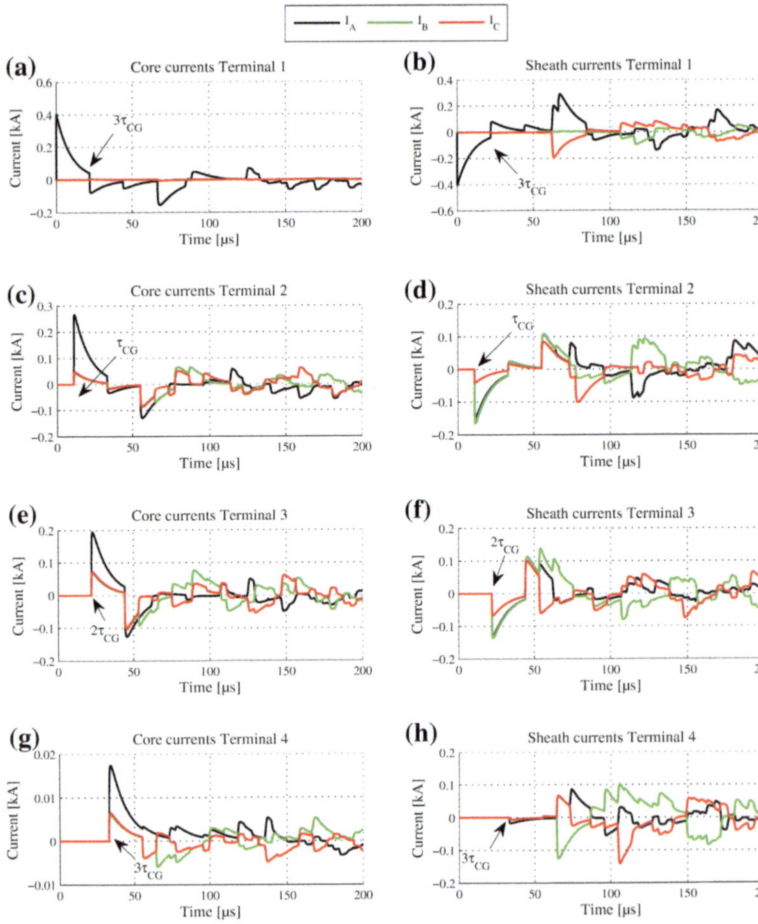

Fig. 6.10 Core and sheath currents at Terminal 1, Terminal 2, Terminal 3 and Terminal 4 after an impulse voltage is applied phase A of a three singe core crossbonded cable

The reason for why no intersheath mode 2 voltage wave is seen at Terminal 2 and 3 lies in the definition of the modal components. From Fig. 6.9, it can be seen that waves arrive on the sheath of cable A and C only. An intersheath mode 1 voltage is from Eq. 6.16 defined as the difference between the voltage existing on these two sheaths. As they are equal, but with opposite signs, an intersheath mode 1 wave exists. An intersheath mode 2 voltage wave is defined as $U_{I2} = -0.5U_{sA} + U_{sB} - 0.5U_{sC}$. As the sheath voltages have different polarity, no intersheath mode 2 voltage exists at Terminal 2 and 3.

Observing Fig. 6.11g and h, it is noticed that no intersheath mode waves are created at Terminal 4 when the first coaxial wave arrives. This is a result of how the 500 Ω termination resistances are connected. From Fig. 6.8, it can be seen that these

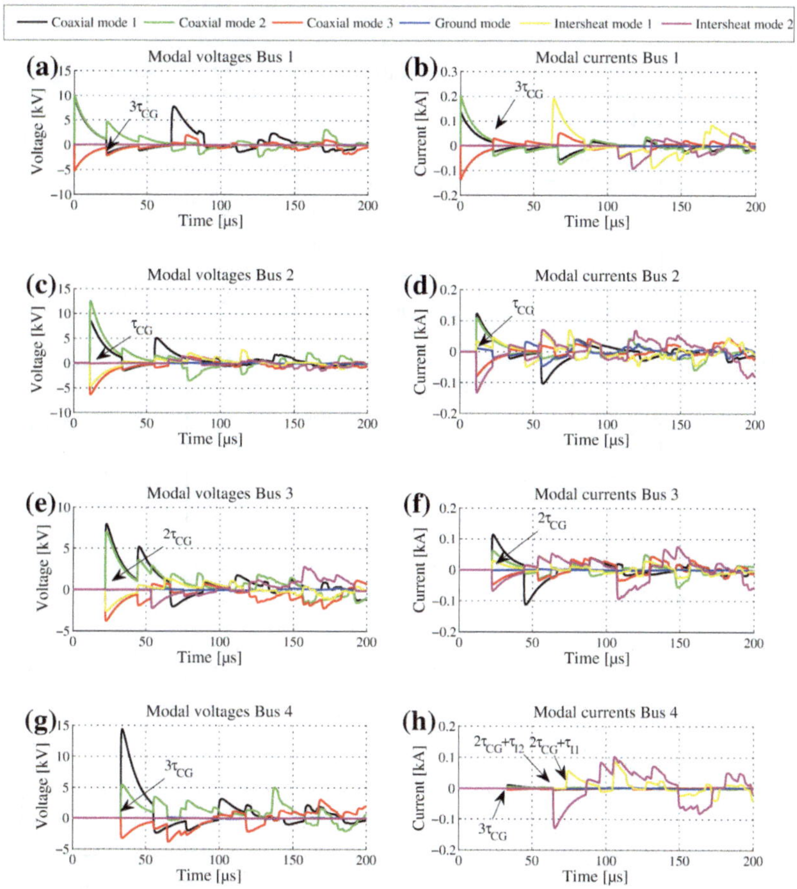

Fig. 6.11 Modal voltages and currents at Terminal 1, Terminal 2, Terminal 3 and Terminal 4 after a 1.2/50 μs voltage impulse is applied phase A of a three single core crossbonded cable

resistances are connected directly between core and sheath (coaxial mode), and thus, no intersheath modes are excited.

The first intersheath current wave seen at Terminal 4 (Fig. 6.11h) is an intersheath mode 2 wave arriving at $2\tau_{CG} + \tau_{I2} = 64\,\mu s$. The first intersheath mode 1 wave arrives at $2\tau_{CG} + \tau_{I1} = 72.7\,\mu s$. These intersheath mode waves are the ones created at Terminal 3 by the coaxial mode wave that arrives at $2\tau_{CG}$. These intersheath mode waves travel from Terminal 3 at their respective velocities until they reach Terminal 4. This is an important finding for fault location purposes: The first intersheath mode wave seen at the FLT has not travelled from the fault location, but was created when the first coaxial mode wave from the fault location met the crossbonding closest to the FLT. An substantial error will thus be made if the intersheath mode velocity is used as input in Eq. 2.1 or 2.2 without compensation.

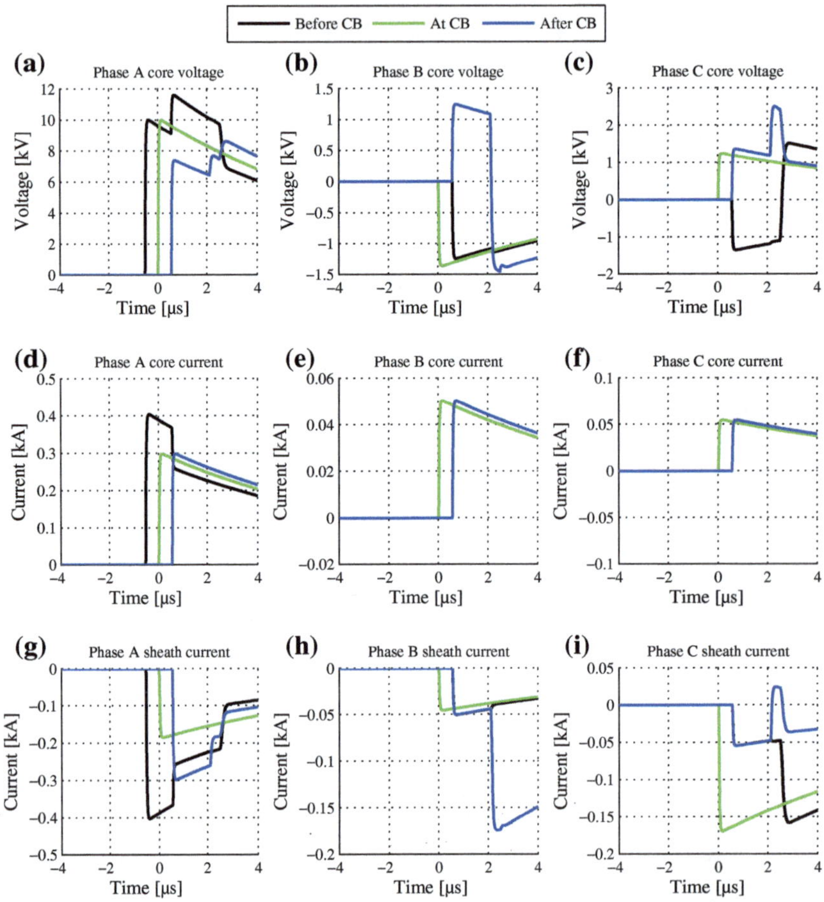

Fig. 6.12 Core and sheath current and core voltage for Phase A, B and C just before, at, and just after a crossbonding

6.2.1 Wave Reflections and Refractions at Crossbondings

To examine the behaviour of the voltage and current waves at the crossbondings in all three phases, the core voltages and currents as well as the sheath currents are obtained just before the crossbonding, at the crossbonding and just after a crossbonding after Phase A is applied an impulse. The time scale is shifted so the wave arrives at the crossbonding at $t = 0$ and the result is shown in Fig. 6.12 where the row show the three phases respectively, and the columns show the core voltage, core current and sheath current.

Figure 6.12a shows that only a part of the incident wave is refracted on cable A at the crossbondings. It is also noticed that voltage waves appear on the core of Phase

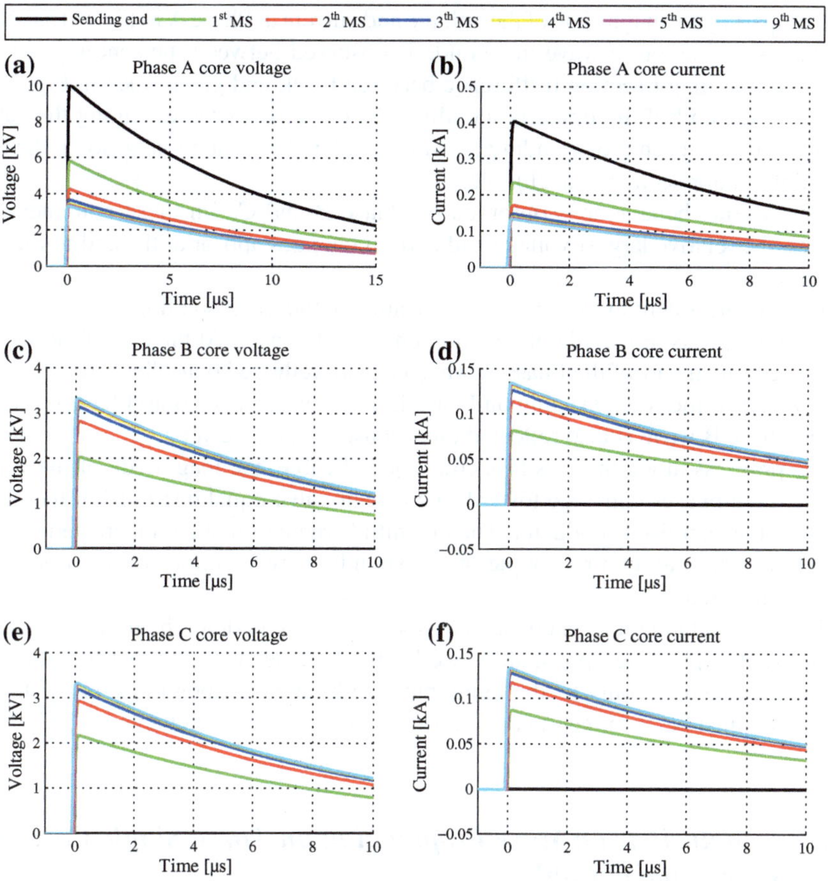

Fig. 6.13 Core voltage and current at the sending end, after major section (MS) 1–5 and after major section 9 on a loss-less cable

B and C at the same instance (green trace in Fig. 6.12b and c). The same is the case for the core and sheath currents.

It can be a problem that a substantial part of an incident fault wave on the faulted phase is not transmitted as that wave that has to be picked up at the FLT has a smaller magnitude. This phenomenon does not occur on OHLs or cables which are not crossbonded and is a specific problem for crossbonded cables. The significance is examined by implementing a loss-less 10 major section crossbonded cable system in PSCAD/EMTDC (section length is of no importance due to the loss-less nature).

The core voltages and currents are recorded on each phase at the end of the first five major sections and at the end of the ninth major section after Phase A is energised at 'the sending end'. The time scale is shifted so the waves at all crossbondings appear at $t = 0$, and the results are shown in Fig. 6.13.

Comparing the magnitude of the voltage and current at Phase A in Fig. 6.13a and b, the largest reduction in wave magnitude is observed between the sending end and after the first major section (difference between black and green trace in Fig. 6.13a and b) at the end of the first crossbonding. As the waves progress along the cable, the reduction at each crossbonding become smaller. The reduction between the fifth and ninth major section is negligible.

The inverse behaviour is observed on Phase B and C. On these non-energised phases, the relative largest voltage and current changes appear at the end of the first major section.

Due to the interconnection of the sheaths at the crossbondings, a part of the incoming wave is refracted onto the non-energised phases. At the end of the second major section, a part of the Phase A wave is again reflected onto Phase B and C, but here, a part of the wave existing on Phase B and C is reflected onto Phase A. A part of the Phase B wave is also reflected onto Phase C and vise versa. Therefore, as the waves progress along the loss-less crossbonded cable, the magnitude of the waves on the three phases converge towards a constant value equal one third of the initial wave on the energised conductor. The magnitude reduction of a fault created wave is therefore largest at the first couple of crossbondings whereafter the effect becomes of less importance.

In case of a fault on an energised crossbonded cable system, the power frequency component is presented pre-fault as well. This together with the fact that the fault wave is reduced not only by damping, but also by the crossbondings, is a potential issue that must be examined in detail. This is done in the next chapter.

6.2.2 Conclusions on Wave Propagation on Three Single Core and Crossbonded Cables

Some important findings for fault location purposes on crossbonded cable system were presented in this chapter.

The first signal change noticed at any location will be due to a coaxial wave and the coaxial wave velocity is not affected by the crossbondings. The main idea of the two-terminal fault location method is to detect only the first wave arriving at each cable end, and the coaxial mode seems therefore ideal as input for this.

Using the intersheath mode waves for the two-terminal method is not straightforward as the first intersheath wave arriving at the fault locator terminal is not directly from the fault location, but was created when a coaxial wave met the crossbonding closest to the terminal. Both the core voltages and currents as well as the sheath currents are of equal interest so far, as they propagate at coaxial velocity with equal behaviour at the crossbondings and along the cable-line. The sheath currents are of special interest as they can be measured using low voltage high bandwidth equipment.

The single ended terminal method is much harder to implement in practice on a crossbonded cable system even when using the coaxial wave type. The first and second waves from the fault location must be used to estimate the fault location. On a solid bonded cable systems, the second waves to reach the fault locator is from the fault assuming no refections from the surrounding grid, whereas on a crossbonded cable system, this is clearly not the case. The performance of the the single and two-terminal methods using the different modal components as input is examined in the next section where actual fault location on crossbonded cables is studied in detail.

References

1. C.K. Jung., J.B. Lee, Fault location algorithm on underground power cable systems using noise cancellation technique. in *Transmission and Distribution Conference and Exposition, 2008*, pp 1–7. IEEE/PES, April 2008
2. L.M. Wedepohl, D.J. Wilcox, Transient analysis of underground power-transmission systems: system-model and wave-propagation characteristics. Proc. Inst. Electr. Eng. **120**(2), 253–260 (1973)
3. L.M. Wedepohl, Application of matrix methods to the solution of travelling-wave phenomena in polyphase systems. Proc. Inst. Electr. Eng. **110**(12), 2200–2212 (1963)
4. L.M. Wedepohl, Electrical characteristics of polyphase transmission systems with special reference to boundary-value calculations at power-line carrier frequencies. Proc. Inst. Electr. Eng. **112**(11), 2103–2112 (1965)
5. A. Ametani, Wave propagation characteristics of cables. IEEE Trans. Power Apparatus Syst. PAS **99**(2), 499–505 (1980)
6. H.W. Dommel, Digital computer solution of electromagnetic transients in single-and multi-phase networks. IEEE Trans. Power Apparatus Syst. PAS **88**(4), 388–399 (1969)
7. A. Ametani, A general formulation of impedance and admittance of cables. IEEE Trans. Power Apparatus Syst. PAS **99**(3), 902–910 (1980)
8. H.M.J.S.P. De Silva, *Accuracy and Stability Improvments in Electromagnetic Simulations of Power Transmission Lines and Cables*. (University of Manitoba, Canada, 2008)
9. L.M. Wedepohl, H.V. Nguyen, D.A. Irwin, Frequency-dependent transformation matrices for untransposed transmission lines using newton-raphson method. IEEE Trans. Power Syst. **11**(3), 1538–1546 (1996)
10. N. Nagaoka, A. Ametani, Transient calculations on crossbonded cables. IEEE Trans. Power Apparatus Syst. PAS **102**(4), 779–787 (1983)
11. Y. Itoh, N. Nagaoka, A. Ametani, Transient analysis of a crossbonded cable system underneath a bridge. IEEE Trans. Power Delivery **5**(2), 527–532 (1990)
12. U.S. Gudmundsdottir, B. Gustavsen, C.L. Bak, W. Wiechowski, Field test and simulation of a 400-kv cross-bonded cable system. IEEE Trans. Power Delivery **26**(3), 1403–1410 (2011)
13. L. Marti, Simulation of transients in underground cables with frequency-dependent modal transformation matrices. IEEE Trans. Power Delivery **3**(3), 1099–1110 (1988)
14. U.S. Gudmundsdottir, *Modeling of Long High Voltage AC Cables in Transmission Systems*, 2nd edn. (Energinet.dk, Denmark, 2010). ISBN 978-87-90707-73-6

Chapter 7
The Use of the Single and Two-Terminal Fault Location Method on Crossbonded Cables

Concluding on the studies so far, the simplicity of the two-terminal fault location method seems beneficial for fault location on crossbonded cables. The instance the first fault wave arrives at the two fault locator terminals is the only variable that needs to be determined for accurate fault location if the wave velocity of the selected mode is known (see Eq. 2.2). It is, however, still of interest to examine the use of the single-terminal fault locator method as no GPS-synchronisation and only half the data acquisition equipment is needed.

Figure 7.1 shows the basic principle for fault location using the two and single-terminal methods on a system with one discontinuity between the faut locator terminals.

As discussed in Sect. 6.2, a coaxial wave will travel between the core and the sheath from the fault location and directly to the fault locator terminals without the velocity being affected by the crossbondings. If the coaxial wave velocity at a given frequency is known and the exact arrival instances τ_{1A} and τ_{1B} defined in Fig. 7.1 can be extracted from the measured signals, then exact fault location is possible.

The single-terminal method relies on detecting both the first and second wave from the fault location arriving at the FLT (see Eq. 2.1). On a system with no discontinuities, the second wave to reach the fault locator is directly from the fault location, whereas on a crossbonded cable system, this is not the case due to the discontinuities in the surge impedance along the cable line. The second reflection τ_{2A}, seen in Fig. 7.1 does not contain the required information for fault location determination, as it was created by the first fault wave at the discontinuity. The time difference between τ_{1A} and τ_{3A} is the correct parameter for the single-terminal method to determine the fault location under the assumption that $l_1 = l_2$. If $l_1 < l_2$, the third wave at Terminal A is from the crossbonded as well.

Another practical issue that complicates fault location using the single-terminal method is the unequal lengths between discontinuities. When, as in the example used in this section, all minor sections are equal in length, all modal wave types will arrive at a multiple of the travelling time of one minor section for the specific mode. If the minor sections are not of equal length, this is not the case. This behaviour is shown

C. F. Jensen, *Online Location of Faults on AC Cables in Underground Transmission Systems*, Springer Theses, DOI: 10.1007/978-3-319-05398-1_7,

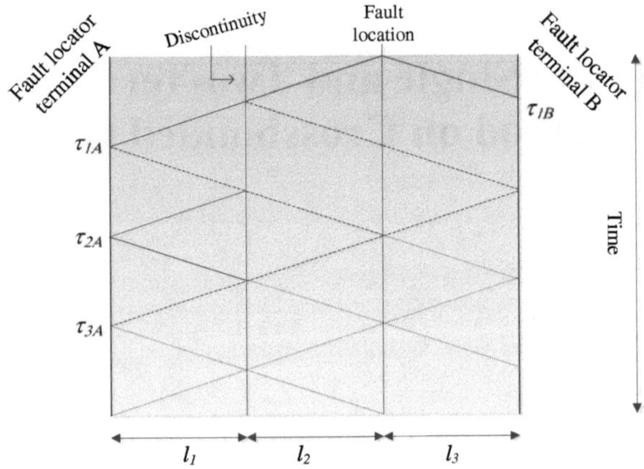

Fig. 7.1 Fault location using coaxial mode waves for single- and two-terminal methodl

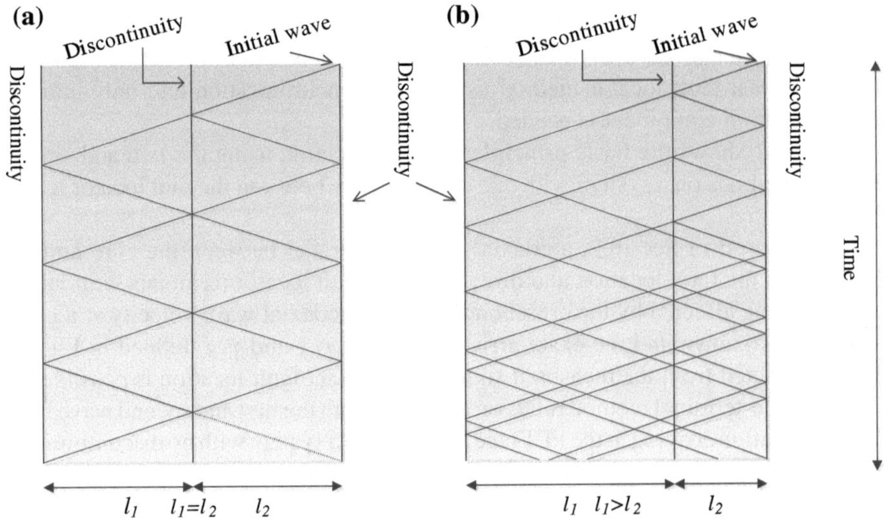

Fig. 7.2 Coaxial wave flow in a cable system with **a** equal lengths between discontinuities and **b** with unequal lengths

in Fig. 7.2b, and for comparison reasons, the propagation in sections of equal lengths is shown in Fig. 7.2a.

Normally, an effort is made to obtain minor sections of equal length due to the balancing of the induced sheath voltage, but the major sections are often not of equal length. Major sectionsof unequal lengths will give rise to the same behaviour

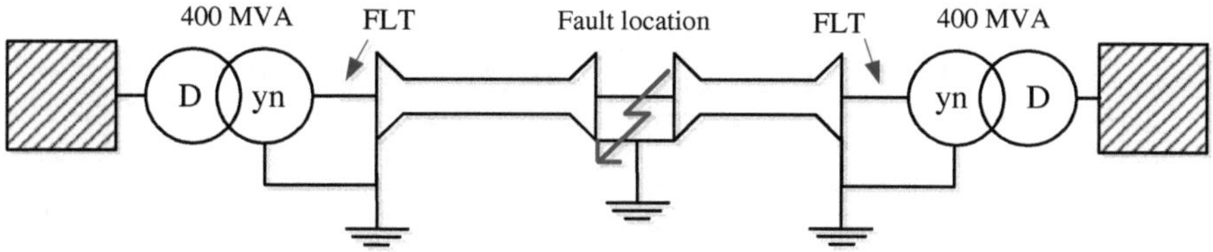

Fig. 7.3 Sketch of case study system used for fault location studies

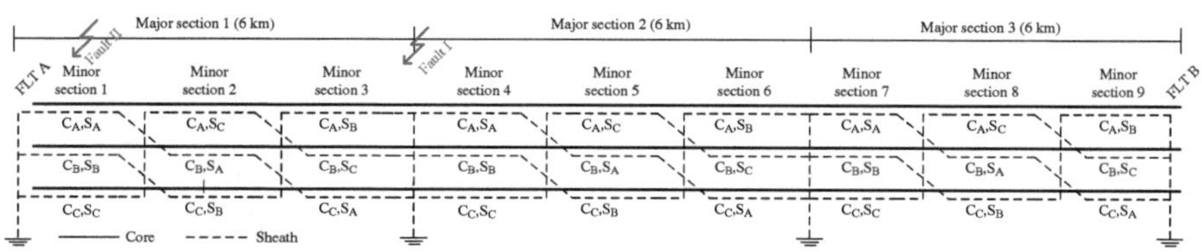

Fig. 7.4 Crossbonded cable with three major sections with fault location I and II marked

as shown in Fig. 7.2b. The waves propagating in the system thus quickly multiply, and therefore, further complications are introduced for using the single-terminal method.

7.1 Fault Location on a Crossbonded Cable System Using Travelling Waves

To examine which fault location methods, which domain and which signal are best suited for fault location on crossbonded cable systems, a case study system is built in PSCAD/EMTDC. The case study 165 kV flat formation cable system described in Sect. 4.1 is studied in this section with the feeder system described in Appendix B. The cable system has three major sections all with lengths of 6 km, and all minor sections are 2 km long. The cable is terminated into a single 400 MVA power transformer at both ends as shown in Fig. 7.3.

The core voltages and currents as well as the sheath currents are recorded at the cable ends (FLT) with a sampling frequency of 10 MHz. The signals are recorded directly at the high voltage terminals; therefore, the influence of the instrument transformers is disregarded. Their effect on the travelling waves is examined in Chap. 8.

After the system has come to steady state, a single phase fault is applied between the core and sheath of phase A at two different fault locations, one at the time. Case Study I involves a fault occurring in the joint 6 km from Terminal A between major section 1 and 2 (Fault I), and Case Study II occurs 1 km from Terminal A in the middle of the first minor section (Fault II). A sketch of the system and the crossbonded scheme is shown in Fig. 7.4.

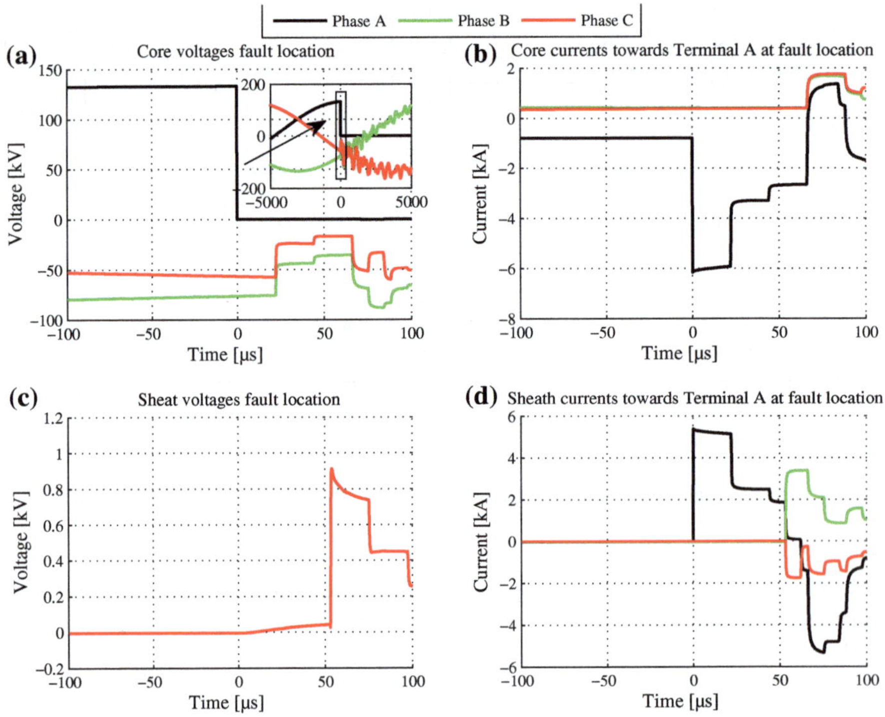

Fig. 7.5 Core voltage, core current and sheath currents at the fault location with current direction towards Terminal A for a phase A fault 6 km from Terminal A on a 18 km crossbonded 165 kV cable system (Fault I)

7.1.1 Case Study I (Fault I)

The system behaviour after a fault is applied between core and sheath of phase A at the end of the first major section is studied first. However, before the fault locator terminal (FLT) voltages and currents are presented, the same quantities at the fault location are shown in Fig. 7.5.

Figure 7.5 shows that at the instance of the fault occurrence, square waves superimposed on the power frequency signal propagate from the fault location in the core voltage and in the core and sheath currents. As already discussed, the sheath provides perfect electromagnetic shielding at high frequencies; thus no coupling exists between the faulted phase and the adjacent cables. No waves are therefore created at the moment of fault in the non-faulted phases.

Which modal components are created at the fault location can be determined by transforming the phase domain quantities seen in Fig. 7.5 to the modal domain. Due to the high frequency content of the phase domain signals at the fault instance, the modal transformation matrices ($\mathbf{Q}^{-1}$ and $\mathbf{S}^{-1}$) evaluated at 1 MHz will give accurate

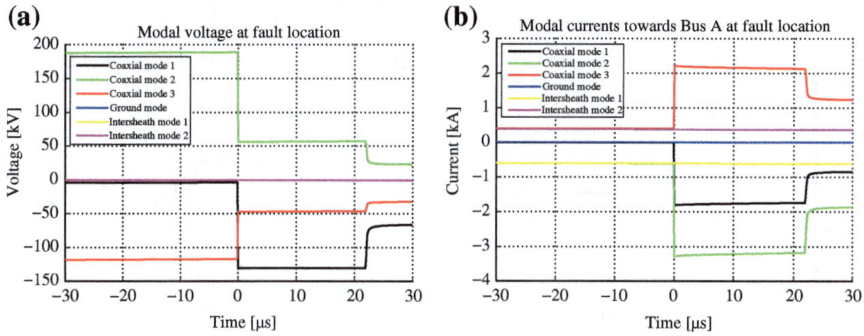

Fig. 7.6 Modal **a** voltages and **b** currents at the fault location with current direction towards Terminal A for a phase A fault 6 km from Terminal A on a 18 km crossbonded 165 kV cable system (Fault I)

results at the fault instance. The modal signals at the fault location are shown in Fig. 7.6.

Figure 7.6 shows that three coaxial voltage and current waves will propagate from the fault location. No change in the intersheath or ground mode waves is observed. This is the same finding as made in Sect. 6.1 where pulse propagation on homogenous cable systems was examined.

In Fig. 7.7, the three-phase core voltages and currents as well as the sheath currents are shown at the fault locator terminals (Terminal A and B).

Even though both square core voltage, core current and sheath current waves propagated from the fault location toward both terminals, Fig. 7.7 shows a clear difference between the core voltage and the core and sheath current signals as the first coaxial wave reaches the terminals. The travel delay τ_{CG} for one 2 km minor section was given in Table 6.2 as 11.03 μs. At $t = 3\tau_{CG} = 33.08$ μs, the first coaxial wave group is seen clearly in the core voltage signal at Terminal A 6 km from the fault location (Fig. 7.7a). Changes are noticed in the core and sheath currents at $3\tau_{CG}$ as well, but the rate of change is much slower. The same is observed at Terminal B at $t = 6\tau_{CG} = 66.16$ μs where the first coaxial wave arrives. The cause of the different behaviour between the voltage and current signals is interesting and is found in the substation's surge impedance.

The substation's surge impedance affecting the coaxial waves will for the example feeder system be dominated by the inductance of the auto transformer and shunt reactor. As no current can penetrate the windings instantaneously, the coaxial wave meets what appears momentarily as an open end, and the amplitude of the incoming wave is doubled. The incoming current wave is reflected at the terminal with a reflection coefficient equal -1 and this wave is added to the incoming wave and effectively reducing the wave to zero [1].

The current in the faulted phase will gradually build up where the sloop of the current will depend on the size of the combined equivalent inductance of the transformer and reactor as well as the coaxial surge impedance of the cable. This shows

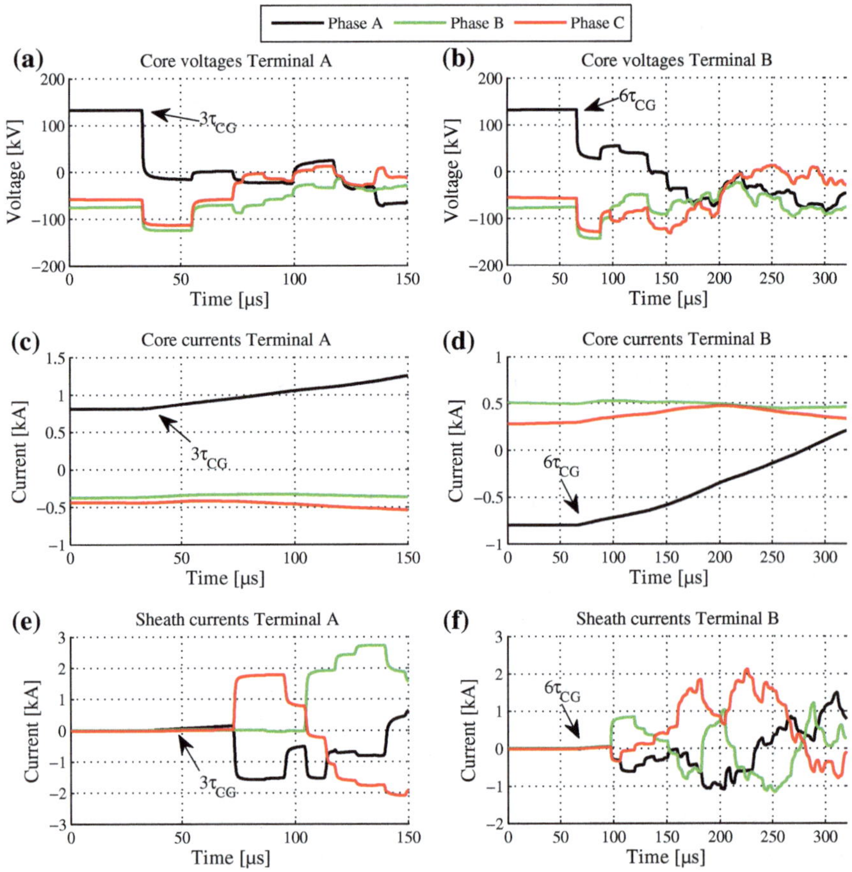

Fig. 7.7 Core voltage, current and metal sheath currents for a Phase A fault 6 km from Terminal A on a 18 km crossbonded 165 kV cable system (Fault I)

the importance of studying the surge impedance of the substation where the fault locator equipment is installed.

In general, the asymptotic coaxial lossless surge impedance of high voltage power cables is between 20 and 50 Ω where the surge impedance for OHLs is around 350–400 Ω. Connecting an additional 165 kV case study crossbonded cable at both Terminal A and B will have a profound impact on the FLT signals. The same fault type as previous is applied at the same location and the core voltage, core current and sheath currents are shown in Fig. 7.8.

The surge impedance governing the behaviour of the coaxial waves is now determined by a combination of the auto transformer, the reactor and the additional crossbonded cable connected at each Terminal. As mentioned before, both the auto transformer and the reactor appear as open ends wherefore the surge impedance of the additional cable will determine the substation's surge impedance the instance the fault wave arrive. As an additional cable of the same type is connected (same coaxial

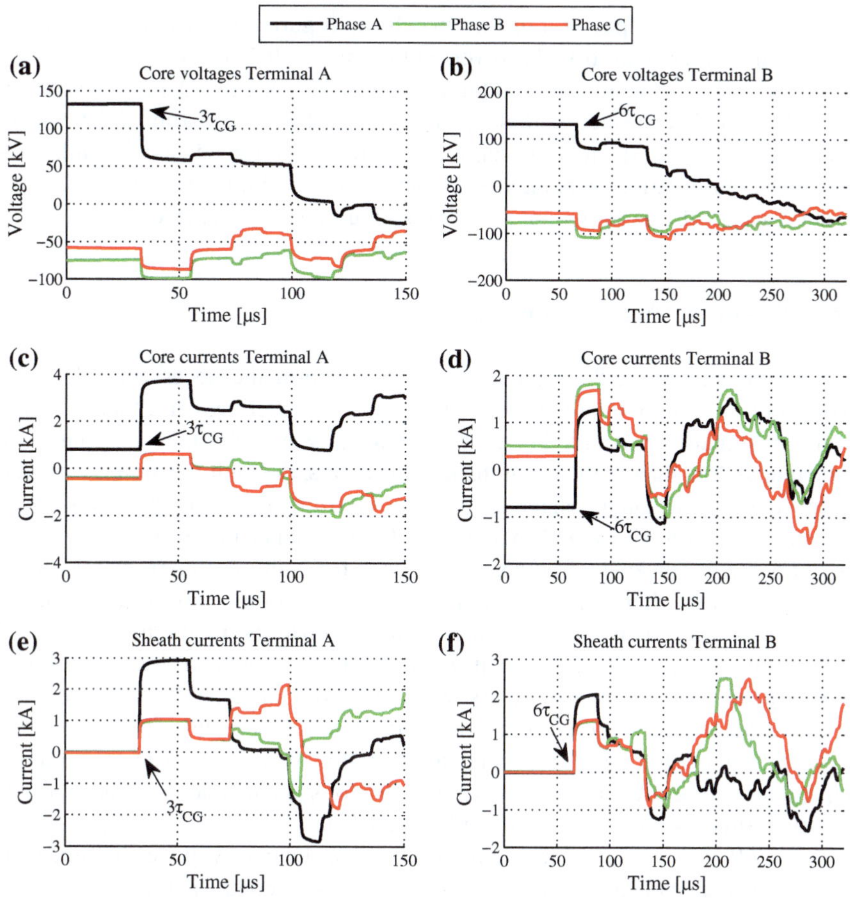

Fig. 7.8 Core voltage, current and metal sheath currents for a phase A fault 6 km from Terminal A on a 18 km crossbonded 165 kV cable system with an additional 6 km crossbonded cable of same type connected at Terminal A and B (Fault I)

surge impedance), the reflection coefficients for all signals become zero and the incident fault waves are transmitted into the additional cable. As more cables or OHLs are connected, the reflection coefficients for the voltage goes towards −1 where the reflection coefficients governing the current go toward 1. For an infinite number of additional lines connected to the substation, the voltage wave is fully reflected and cannot be used for fault location as it is the case for the current waves when a single transformer/single line configuration is used. The effect of the substation is studied in detail in Chap. 8.

The findings here indicate that the signal type used as an input for a fault locator unit is system dependent leading to a benefit of using either core currents or sheath currents for stations with many lines and voltage waves of stations with only the monitored cable and a power transformer. The configuration where a single line per power transformer is used is, however, not common. It is used in some cases for connecting offshore wind farms to the main grid, or else, the Danish transmission grid will be highly meshed with several cables at the same Terminal. The special offshore connection type and the full effect of the surge impedance dependence on substation configuration are studied in further detail later due to monitoring especially the submarine cable is important seen from a fault location perspective.

Estimation of the Fault Location

It is common to use a decoupled domain to perform the necessary calculations involved in estimating the fault location with use of travelling wave theory. Voltage or current signals are recorded in the time domain where after they are transformed to the modal domain and by use of the modal signals, the fault location is estimated. However, at high frequencies, the transformation from the time to the modal domain and vise versa is done by multiplications of real-number matrices. As the coaxial waves are the fastest at any frequency of interest for fault location purposes, the first voltage or current change noticed in the time domain in the case of fault is caused by a coaxial wave. Thus, there is no need to calculate the modal transform of the time domain signals before the fault location can be estimated. Using the coaxial wave velocity and the fact that the first wave noticed in the time domain is coaxial of nature, makes fault location directly in the time domain possible. This is under the assumption that a frequency independent coaxial wave velocity determined at high frequencies can be used as input to the fault locator system under all conditions. The validity of this assumption is examined in Sect. 8.1.

Figure 7.9 shows a high time resolution plot of the core voltages at Terminal A and B at the expected arrival instance of the first and second wave directly from the fault location for the system with only the monitored cable connected to the terminal.

Extracting the arrival instance of the first coaxial wave group from the time domain signals using a simple visual inspection results in a fault wave arrival instance of 32.9 μs at Terminal A and 65.9 μs at Terminal B (Fig. 7.9a and b). Using Eq. 2.2, with the coaxial wave velocity given in Table 6.1, the fault location is determined as 6.011 km; an error of 11 m. If the core and sheath currents are studied on the system with an additional cable at each terminal (Fig. 7.8), the same accuracy is obtained. The same fault location information is available in all signals, thus, the input signal can be freely chosen depending on the system's configuration. Also, the type of input signal in the two cable ends needs not to be the same.

The sheath currents are normally not monitored for fault location purposes. This possibility is available for cables only where a low voltage high bandwidth current transducer can be installed for each cable. As shown, coaxial current waves travel on the sheath conductors, and these signals are therefore equally beneficial for use as input for a fault locator system. Figure 7.10 shows that it is quite easy to mount such a current transducer on a real-life cable termination.

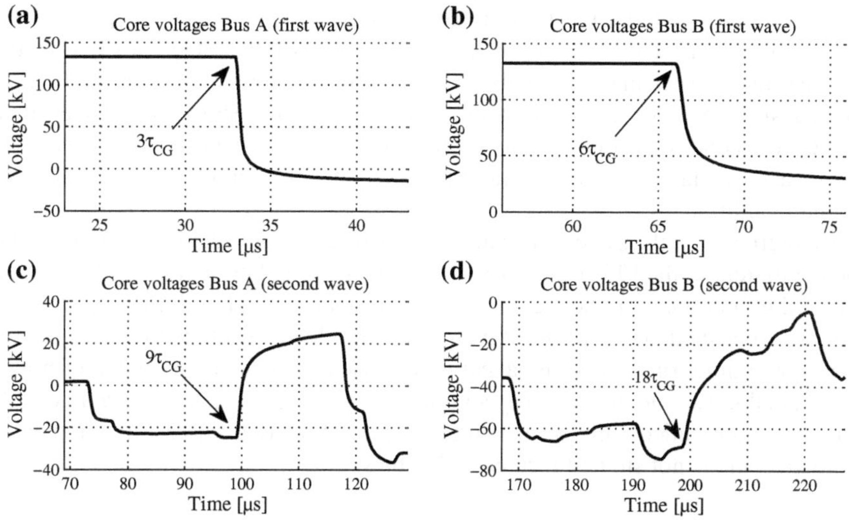

Fig. 7.9 High resolution time zoom of core voltage for a phase A fault 6 km from Terminal A on a 18 km crossbonded 165 kV cable system (Fault I) at the instance of the first and second fault surge arrival instance

Fig. 7.10 Extraction of cable sheath at cable termination seen from **a** a distance and **b** close up

One issue is, however, important; when a cable is energised, the breaker at one end is closed first whereafter the closure of the breaker at the other end follows. When only one breaker is closed, only the voltage is monitored as no current can flow. Therefore, a core or sheath current based fault locator method will not work when cables are being energised.

The second wave to arrive from the fault location is harder to determine in a real-life fault location situation as shown in Fig. 7.10c and d. As shown in Fig. 7.5, a voltage collapse occurs in the phase voltage of the faulted phase at the instance of fault. This wave will travel to the FLT where it is fully reflected due to the inductive nature of the substation's coaxial voltage surge impedance. At the fault location, the voltage reflection coefficient is -1 due to the short between the core and sheath, and the voltage is reflected back as a positive voltage steep. The second coaxial wave from the fault location will arrive at Terminal A at $\tau_{2CG} = 3 \cdot 3\tau_{CG} = 99.3\,\mu s$ and at Terminal B at $\tau_{2CG} = 3 \cdot 6\tau_{CG} = 198.6\,\mu s$. A rather steep positive voltage change is indeed seen in both signals at these instances. However, observing Fig. 7.7c and d, positive steep voltage changes are seen in the signals recorded at both terminals before the second wave arrives. This shows that it will be very problematic to identify the correct second wave from the fault location on a crossbonded system even with optimal condition.

Theoretically, if the system with an additional cable connected is used, the reflection coefficient at the FLT is zero for all signal types and no wave is reflected back toward the fault location; hence, the single-terminal method will not work. If more lines are connected, a part of the wave is indeed reflected back, but reflections from the crossbondings on the additional cable will also contribute to the signal profiles at he FLT, making single-ended fault location even more problematic to use. In conclusion, the single-ended fault location method cannot be used for crossbonded cables in practice and is not studied further in this thesis.

The sheath Current Method for Single Line/Single Transformer Configurations

As the sheath currents are easily accessible on real-life systems, the use of these on single line/single transformer configurations as shown in Fig. 7.3 is examined in more details.

The sheath currents shown in Fig. 7.7e and f show some interesting behaviour.

The initial core and sheath current waves are fully reflected back towards the fault location at the auto transformer and can therefore not be used for fault location as discussed. At the arrival instance of the first coaxial wave, the sheath current of the faulted phase increases slowly as expected. However, at 72.8 and 97.3 μs in Fig. 7.7e and f, steep signal changes in the sheath currents are observed at Terminal A and B. These signals are not of coaxial nature, but are caused by intersheath waves arriving at the FLT. It is interesting to study whether these signals can be used for fault location as an options different from using voltage signal.

As intersheath mode currents are defined as currents flowing between the different sheath circuits (see Fig. 6.3), the intersheath reflection coefficient at the FLT is determined by the bonding of the sheaths. For an intersheath current wave, the reflection coefficient is therefore 1 at the FLT resulting in steep current changes as the intersheath mode waves arrive.

The first intersheath mode wave seen at the FLT has not traveled directly from the fault location as none of these waves are created at the fault location (see Fig. 7.6b).

As shown in Sect. 6.2, intersheath mode waves are created when a coaxial type of wave meets a crossbonding. The first intersheath mode wave arriving at the FLT is, due to this, not travelling directly from the fault location, but is created by a coaxial wave arriving at the crossbonding closest to the FLT. The first intersheath mode wave at the FLT does indeed contain the necessary information about the fault location and by compensating for the slower travelling time in the two minor sections closest to the FLT, fault location can possibly be carried out if the intersheath mode waves

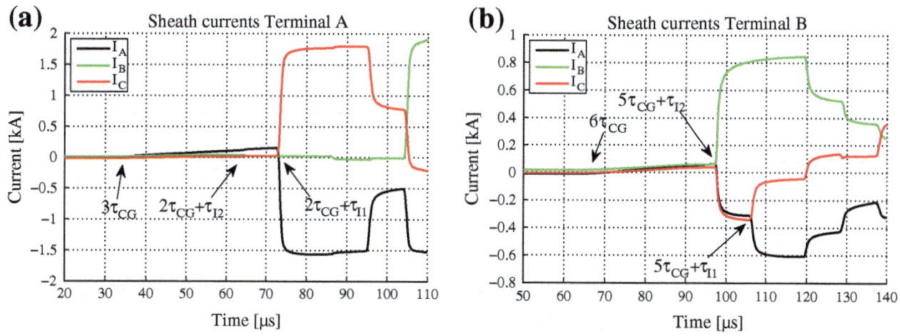

Fig. 7.11 Sheath currents for a phase A fault 6 km from Terminal A on a 18 km crossbonded 165 kV cable system measured at **a** Terminal A and **b** Terminal B

can be extracted from the sheath current signals. A high time resolution zoom of the sheath currents at Terminal A are shown in Fig. 7.11.

The first wave to arrive at Terminal A is, as mentioned, caused by a coaxial wave group ($3\tau_{CG} = 33.1\,\mu$s). This coaxial wave met the crossbonding closest to the Terminal A at $2\tau_{CG} = 22.02\,\mu$ where intersheath mode waves were created. By using the intersheath mode travelling times presented in Table 6.2, the Intersheath mode 2 wave is expected at $2\tau_{CG} + \tau_{I2} = 64.1\,\mu$s and Intersheath mode 1 wave at $2\tau_{CG} + \tau_{I1} = 72.7\,\mu$s. From Fig. 7.11a, it is clearly seen that the current profile is dominated by an Intersheath mode 1 wave.

At Terminal B, the result is different. The Intersheath mode 2 wave is expected at $5\tau_{CG} + \tau_{I2} = 97.15\,\mu$s where intersheath mode 1 should arrive at $2\tau_{CG} + \tau_{I1} = 105.7\,\mu$s. At this terminal, the sheath current profile is dominated by an Intersheath mode 2 wave.

As the two intersheath wave types travel with different velocities for a cable system in flat formation, it is necessary to know which type of wave has arrived at the FLT so the correct compensation can be carried out.

Identifying the intersheath wave type can be done in several ways. The modal signals can be calculated using the three core and the three sheath currents. This is not realistic in a real-life system as all signals must to be acquired at a very high sampling frequency.

Figure 6.1 shows how the modes are defined. An Intersheath mode 1 current for a system laid in flat formation is defined as 1 pu current flowing into phase A and returning in phase C; phase B is not involved in this mode. Intersheath mode 2 is defined as a current flowing into phase B and returning equally divided in phase A and C. At Terminal A (Fig. 7.11a, where an Intersheath mode 1 wave was identified, only a current change in phase A and C is noticed. At Terminal B, the intersheath mode wave affects all three-phase currents, thus, an Intersheath mode 2 wave is identified.

Why the different modes are excited at the two cable ends can be realised by observing the crossbonded cable system shown in Fig. 7.12 where a fault is applied in the joint between the two major sections one phase at the time.

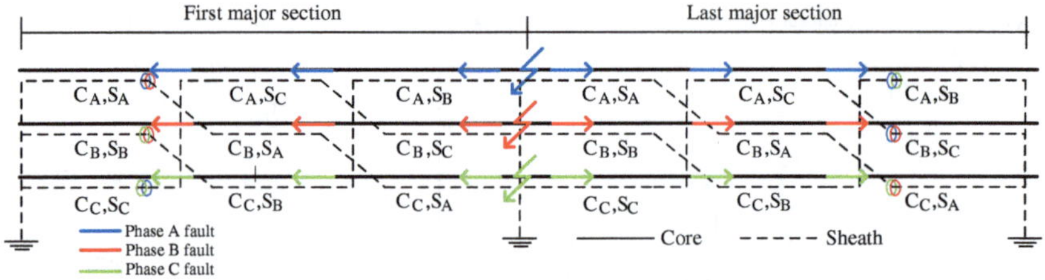

Fig. 7.12 Sheaths affected by a current flowing in the core of phase A, B and C for a fault after the first major section on a crossbonded cable system

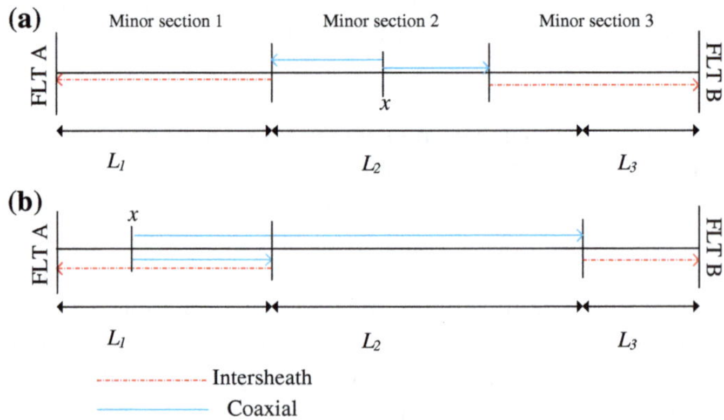

Fig. 7.13 Wave propagation in case of a fault **a** in another than the two minor sections closest to the FLT and **b** for a fault in the first minor section

When a wave travels towards Terminal A and meets the crossbonding closest to the FLT, two sheaths are electrically connected at this point. In case of a Phase A fault, this is sheath A and C. These are the two sheaths involved in Intersheath mode 1 and the one mode that dominates the sheath current signals at Terminal A (see Fig. 7.11a). At Terminal B, the sheath of cable A and B are directly electrically connected. This creates a dominating Intersheath mode 2 wave as seen in Fig. 7.11b. This means that the dominating intersheath mode wave to arrive at the FLT is dependent on in which phase the fault occurs.

In case of a cable fault, the protection relays determines the faulted phase. Furthermore, the crossbonding scheme, and thereby which intersheath mode wave is excited, is known in advance. The method seems feasible, however, it contains one problem. It was shown in Fig. 7.6 that no intersheath mode waves are created at the fault location when the fault occurs. This means that if the fault occurs in one of the minor sections closest to the FLT, the wave propagation route becomes different than if the fault occurs in any of the other minor sections. The wave propagation, in case of a fault at the location x, is shown in Fig. 7.13 where minor section 2 can represent any number of sections.

For the case shown in Fig. 7.13a, the time difference $\tau_A - \tau_B$ used as input to the two-terminal fault location methods is calculated as:

$$\tau_{AB,1} = \left(\frac{x}{v_C} + \frac{l_1}{v_{In,A}}\right) - \left(\frac{l_2 - x}{v_C} + \frac{l_3}{v_{In,B}}\right) \tag{7.1}$$

$$= \frac{2x}{v_C} + \frac{l_1}{v_{In,A}} - \frac{l_2}{v_C} + \frac{l_3}{v_{In,B}} \tag{7.2}$$

where τ_A and τ_B are the dominating intersheath mode arrival instances at Terminal A and B respectively, x is the fault location seen from the end of the first minor section, $v_{In,N}$ is the velocity of an intersheath mode wave of type n on the minor section closest to Terminal N, and v_{CG} is the coaxial group mode velocity.

If the fault location x is isolated and the length of the first minor section seen from Terminal A is added, the fault location with reference to Terminal A is determined. For the case shown in Fig. 7.13b, the time difference $\tau_A - \tau_B$ is found as:

$$\tau_{AB,2} = \left(\frac{l_1 - x}{v_C} + \frac{l_1}{v_{In,A}}\right) - \left(\frac{l_1 - x}{v_C} + \frac{l_2}{v_C} + \frac{l_3}{v_{In,B}}\right) \tag{7.3}$$

$$= \frac{l_1}{v_{In,A}} - \frac{l_2}{v_C} - \frac{l_3}{v_{In,B}} \tag{7.4}$$

The time difference becomes independent of the fault location x, and, as a result, the fault location cannot be determined if the fault occurs in any of the two minor sections closest to the terminals. This renders the method unusable as any fault location scheme should cover the entire cable. As a result, voltage signals must therefore be measured when single tranformer/single lines configurations are used.

7.1.2 Case Study II (Fault II)

Fault II represents a fault occurring between the core and sheath of Phase A 1 km from Terminal A in the middle of minor section 1. As shown, the sheath current method will not provide an estimate of the fault location in this case as the time difference $\tau_A - \tau_B$ becomes a constant.

In Fig. 7.14, the three-phase core voltages and currents as well as the sheath currents are shown at Terminal A and B.

The core voltage of the faulted phase (phase A) at Terminal A is different from the one recorded at Terminal B. One dominating frequency is seen with large magnitude changes as the wave arrives. The coaxial voltage wave will propagate from the fault location to the terminal where it is fully reflected. When it reaches back to the fault location, it is reflected with a reflection coefficient equal -1. The coaxial wave must travel to the fault location twice before it obtains the same polarity as the first incoming wave. The frequency of the oscillation is therefore determined as [2].

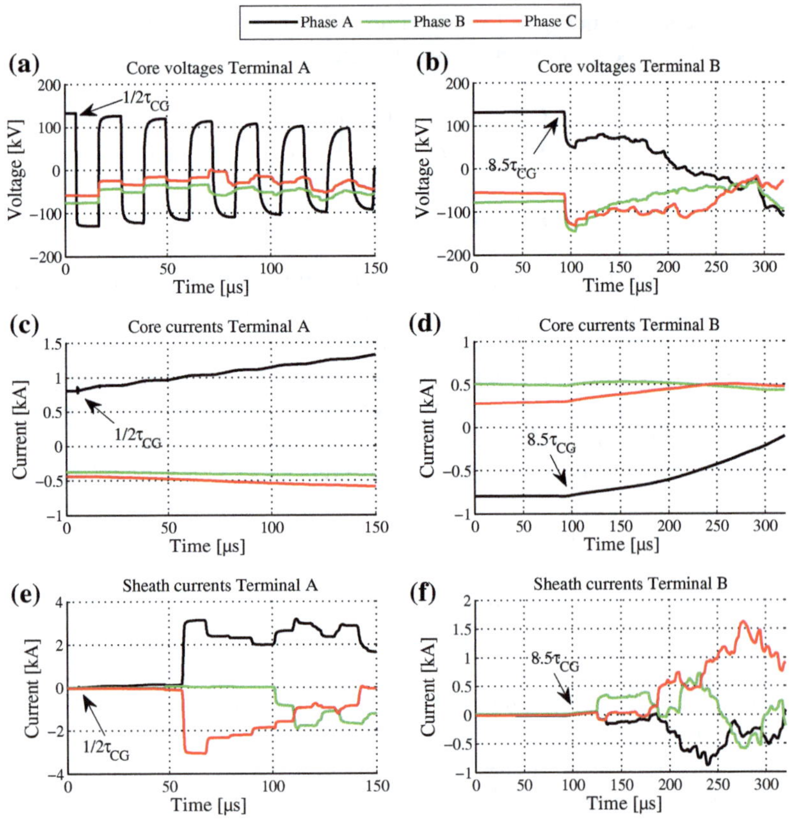

Fig. 7.14 Core voltage, current and sheath currents at the fault location with current direction towards Terminal A for a phase A fault 1 km from Terminal A on a 18 km crossbonded 165 kV cable system (Fault II)

$$f_i = \frac{v_C}{4l_{F1}} \tag{7.5}$$

where v_C is the velocity of the coaxial mode, l_{F1} is the line length from Terminal A.

The relation given in Eq. 7.5 can be used to estimate the fault location if the main frequency component of the signal can be extracted [3–5]. This is, contrary to the current sheath method, only possible if the fault occurs in the minor section closest to the FLT. With the same argumentation as given for the sheath current method, this method is not practically applicable on crossbonded cables.

The two-terminal method with voltage signals as input will give accurate results in this case as well. At Terminal A, the first wave is determined to arrive at 5.4 μ and at Terminal B, at 93.5 μ. Using Eq. 2.2, the fault location is estimated as 1.011—an error of 11 m.

7.1.3 Conclusions on the Use of the Single and Two-Terminal Fault Location Methods on Crossbonded Cables

The two-terminal fault location method using either voltage or current signals analysed directly in the time domain is well suited for faut location on crossbonded cables. Whether to use core voltage, core currents or sheath currents as input is system dependent. For a single transformer, single cable configuration type of connecting voltage signals are preferable where, as the number of lines increases, core and sheath currents tend to be the better choice. The single-terminal is not practically applicable in a transmission system based on crossbonded cables. Reflections from the many crossbondings makes extraction of the second wave from the fault location very difficult wherefore the methods cannot be used.

All conclusions conducted in this section are based on high voltage side signals. On real-life power systems, the high voltage signals are represented in the secondary of measuring transformers. How these affect the transients in the system is of importance and is described in the next chapter. Furthermore, which cable and cable system parameters affect the accuracy of the method must be examined. In this section just presented, only short cables or cables with no attenuation were studied.

References

1. A. Greenwood, *Electrical Transients in Power Systems*, 2nd edn. (Wiley, New York, 1991), ISBN 0-471-62058-0
2. H.W. Dommel, *EMTP Theory Book*, 2nd edn. (Microtran Power System Analysis Corporation, Vancouver, 1996)
3. M. Paolone, A. Borghetti, C.A. Nucci, An automatic system to locate phase-to-ground faults in medium voltage cable networks based on the wavelet analysis of high-frequency signals. in *IEEE Trondheim PowerTech 2011*, pp. 1–7 (2011)
4. A. Borghetti, M. Bosetti, C.A. Nucci, M. Paolone, A. Abur, Integrated use of time-frequency wavelet decompositions for fault location in distribution networks: theory and experimental validation. IEEE Trans. Power Delivery **25**(4), 3139–3146 (2010)
5. A. Borghetti, M. Bosetti, M. Di Silvestro, C.A. Nucci, M. Paolone, Continuous-wavelet transform for fault location in distribution power networks: definition of mother wavelets inferred from fault originated transients. IEEE Trans. Power Syst. **23**(2), 380–388 (2008)

Chapter 8
Parameters Influencing a Two-Terminal Fault Location Method for Fault Location on Crossbonded Cables

Some power system components, cable and cable system parameters will influence the characteristics of the first fault wave seen at the FLTs and thereby influence the ability to extract the necessary information for accurate fault location. In this section, the most important parameters are examined.

- The dispersive media effect and cable length
- Busbar surge impedance
- Fault wave reflection and refraction
- Fault inception angle
- Fault arc resistance
- Sensitivity of the coaxial modal velocity on cable and cable system parameters
- Determination of the modal velocities
- Measuring transformers
- Fault locator sampling frequency.

8.1 The Dispersive Media Effect and Cable Length

As shown in Fig. 6.2, different frequency components will experience different levels of damping; for coaxial waves, the highest frequency components are damped the most. The high frequency content in a signal makes the most accurate fault location possible as the exact arrival instance can be determined more precisely. This can be a problem for longer cables as the total damping becomes higher.

Figure 6.2b shows that different frequency components travel with different velocity (dispersive media effect). An ideal square shaped fault wave created at the fault location contains all frequency components. In general, as the frequency becomes lower so does the velocity, and, as a result the square wave spreads out and changes shape as it travels. This effect combined with the fact that high frequency components are damped the most may make the use of a constant coaxial wave velocity as input to a fault location unit problematic. However, as shown in Sect. 6.1, the coaxial wave

C. F. Jensen, *Online Location of Faults on AC Cables in Underground Transmission Systems*, Springer Theses, DOI: 10.1007/978-3-319-05398-1_8,

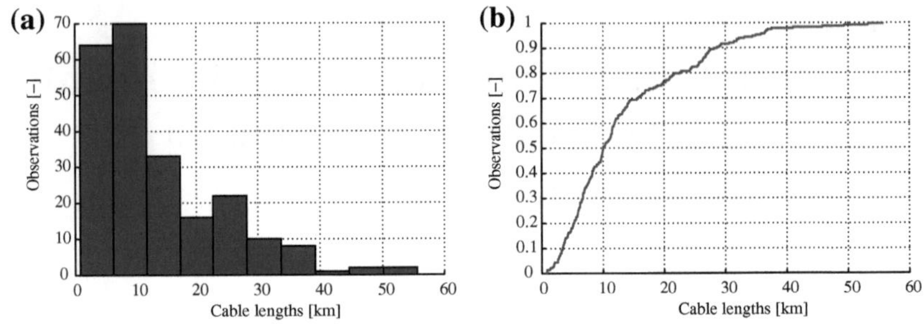

Fig. 8.1 **a** Histogram and **b** empirical cumulative distribution function of planned cable line lengths in the Danish transmission grid in 2030

velocity tends to become frequency independent above 10 kHz, so the total effect on the ability to determine the fault location is interesting to examine. First, however, a histogram and an empirical cumulative distribution function of planned 132 and 150 kV cables in the future Danish transmission grid are shown in Fig. 8.1.

The figures show that approximately 50 % of all cable lines are less than 10 km and 77 % are below 20 km. Few links for connecting existing and planned offshore wind parks are not included in the data set. These links can be of longer length, but most often consist of a crossbonded land cable part and a submarine cable part. For instance, the length of the electrical connection to the Horns Reef 2 offshore platform is 100 km, and the connection to the 400 MW Anholt offshore wind park is 84 km. Fault location on these special systems is examined in detail in Sect. 9.4, but as an introductory study, cables with a lengths up to 100 km are examined in this section as well.

8.1.1 Wave Velocity as Function of Signal Frequency Content

The attenuation of the coaxial modal wave is strongly affected by the resistivities of the core and sheath [1, 2]. Either copper or aluminium is used for the core on high voltage cables. The core conductor is often stranded in order to reduce the influence of the skin effect and to ensure flexibility of the cable. The sheath of a land cable is either a copper or aluminium layer or made from a combination of these. Often a wound wire sheath combined with a solid layer is used. The sheath of a submarine cable is made from lead and acts as a radial water barrier. All layers in the cable model implemented in PSCAD/EMTDC and used for the modal decomposition are represented as solid layers of one material only. Methods for adjusting the parameters are presented in [3, 4].

As the high frequencies in the fault waves are damped, the wave will appear as travelling slower. Because the attenuation is dependent on the resistivities of the core and sheath, it is expected that the error, (if a constant velocity is assumed) increases if material with poorer conductivities are used. Four theoretically 100 km cables with no crossbondings are implemented in PSCAD/EMTDC. In all cases, the use of solid conductors is assumed. The material of the core and sheath are changed to examine the effect of the frequency dependent velocity. The first case is a loss-less cable with infinite electrical conductivity of the core and sheath ($\rho_c = \rho_s = 0$)—this case is used as a reference. The second case is with copper core and sheath ($\rho_c = \rho_s = 1.724e^{-8}\ \Omega$m). A cable with aluminium core and a composite aluminium and copper sheath is used as case three. In case three, an equal distribution of the sheath material is assumed ($\rho_c = 2.8e^{-8}\ \Omega$m and $\rho_s = 2.262e^{-8}\ \Omega$m). A more complete method to take into account the composite sheath is given in [4]. The forth case is a submarine cable with aluminium core and lead sheath ($\rho_c = 2.8e^{-8}\ \Omega$m and $\rho_s = 2.2e^{-7}\ \Omega$m).

A high frequency wave is applied between core and sheath of phase A, and the core voltage at the sending end (0 km), 25, 50, 75 and 100 km is simulated. A coaxial wave velocity evaluated at 1 Mhz ($v_{c,1\text{MHz}}$) is used to calculate the expected travelling delay per 25 km (τ_{25km}). τ_{25km} is subtracted from the time vector for the signal recorded at 25 km, $2\tau_{25km}$ is subtracted from the time vector for the signal at 50 km and so on. For a constant wave velocity of $v_{c,1\text{MHz}}$, all waves should thus appear at $t = 0$. The result is seen in Fig. 8.2.

The results show that the effect of lower wave velocity as function of frequency will cause poorer fault location possibility as higher resistivity materials and longer cables are used. Especially on longer submarine cables (Fig. 8.2d), the effect is profound. Furthermore, the figures show that the exact arrival instance becomes harder to determine exactly as the high frequency components are damped out. Figure 8.3 shows the fault location estimation error as function of arrival instance estimation error at two different coaxial modal velocities.

The frequency dependent wave velocity is not an issue that has to be taken into consideration for fault location on most cables in the future Danish grid. Using a cable with an aluminium core and a combined copper and aluminum sheath, the error introduced by assuming a constant coaxial velocity is 6, 12 and 22 m for cable lengths of 25, 50 and 75 km respectively. As many as 98.6 % of all Danish lines will have length less than 50 km if the grid is constructed as currently planned.

8.2 Busbar Surge Impedance

Previously in this thesis, is has been assumed that the influence of the substation's busbars could be ignored. The construction of the substation can have an influence on the signals measured by the measuring transformers and is examined through an example in this section.

The small outdoor Danish 150 kV substation Nibstrup is used as a case study in this section. The station is situated in the northern part of Jutland and has two main 150 kV cable lines from Hvorupgård and Bredkær, respectively (see Fig. 1.1 for location of the station). Two 100 MVA 150/60 kV transformers and a 40 MVAr shunt reactor are installed at the station. A single line diagram of the station is shown in Fig. 8.4.

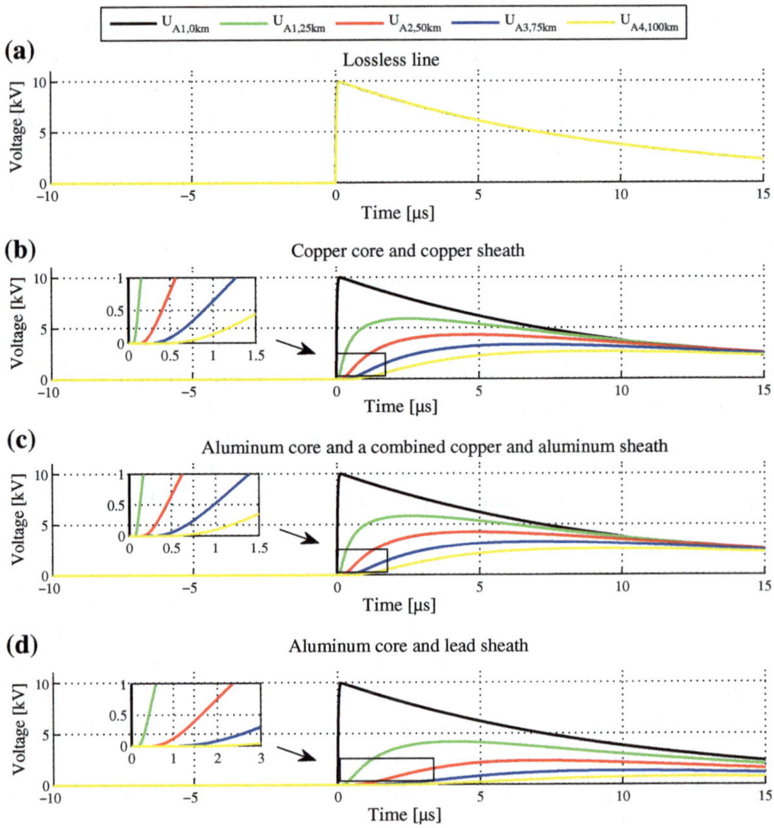

Fig. 8.2 Phase A voltage at the sending end (0 km), 25, 50, 75 and 100 km after an impulse between core and sheath of phase A is applied

Fig. 8.3 Fault location estimation error as function of arrival instance estimation error at two different coaxial modal velocities

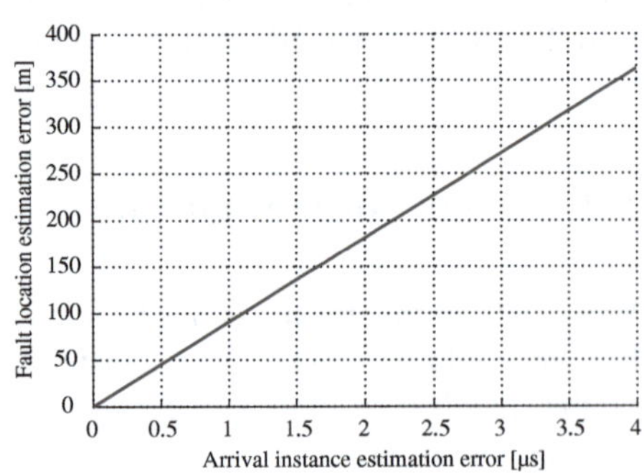

In case of a fault in the cable to Hvorupgård (marked in the figure), voltage and current fault waves are created at the fault location and will travel towards station Nibstrup. As the waves reach the substation, they are guided to the busbar and they

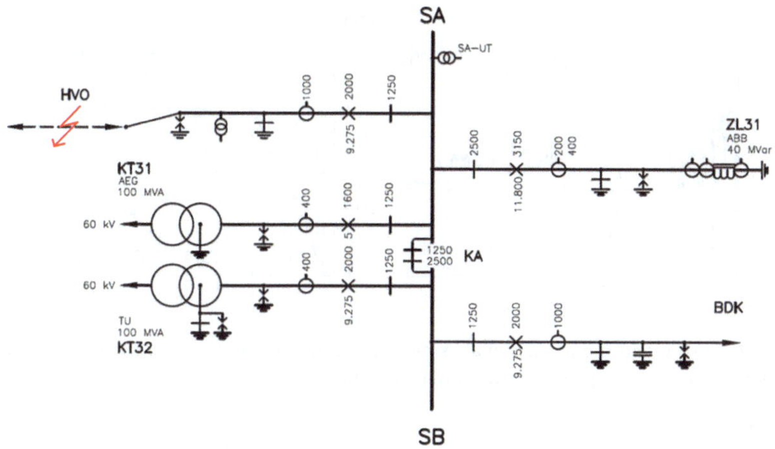

Fig. 8.4 Single line diagram of station Nibstrup

meet a surge impedance around 350 Ω. The result is a wave reflection that gives rise to a voltage of 1.85 pu and a current of 0.15 pu relative to the incoming wave. These waves continue to travel along the busbar until they reach the first discontinuity. Here, the waves are reflected and refracted according to the relationship between the surge impedances at the specific location. At the measuring location, the voltage and current fault waves are still 1.85 and 0.15 pu. The waves from the discontinuities in the station start to arrive back at the measuring location. They meet a reflection coefficient equal, but with opposite sign compared to the one the initial wave travelling from the cable into the busbar met. After all internal substation's waves have settled, the magnitude of the voltage and current fault waves is determined alone by the additional lines connected to the station assuming no current can penetrate the transformer windings during the time it takes the internal substation waves to settle. The same value is theoretically obtained instantaneously if the busbar had no length.

Due to the many discontinuities even in small substations, the number of travelling waves quickly multiplies. To examine the behaviour of the internal substation's waves, Substation Nibstrup is implemented in PSCAD/EMTDC where the busbars are modelled using the Bergeron line model given the surge impedance and busbar length as input parameters. The voltage and current transformers are ignored and the transformers and the reactor are modelled using the standard power frequency components available in PSCAD/EMTDC with a Π-network of capacitors representing the first high frequency response.

A voltage wave is applied to the busbar through a surge impedance of 24.8 Ω representing the 165 kV case study cable. The current and voltage are normalised to the value they obtain after all internal substation's waves have settled. Figure 8.5 shows the response at the measuring location of the cable to Hvorupgård together with a reference representing a busbar with no length.

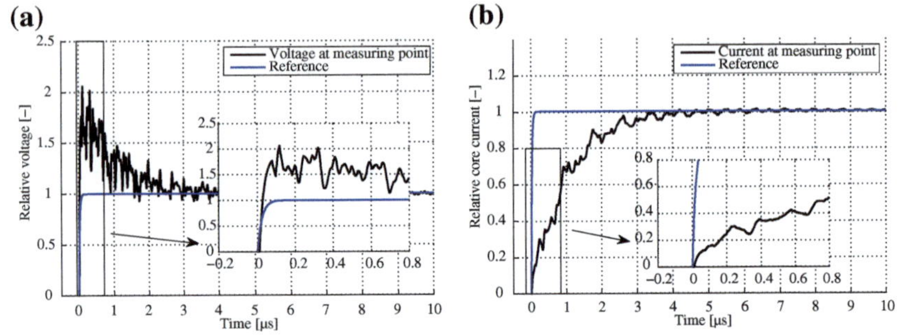

Fig. 8.5 **a** Voltage and **b** current signals simulated at the instrument transformer location at the cable connection to Hvorupgård in substation Nibstrup

Fig. 8.6 **a** GIS switch gear at Vester Hassing and **b** GIS cable termination [7]

Figure 8.5 shows the result of the very different surge impedances between the cable and busbar. The voltage is shortly increased and the current will build up over time by the added reflections in the measuring point. Substation Nibstrup will be one of the smallest in the future Danish grid and as the physical size of the substations increases, the more influence it will have on the fault transients. The substations influence on the voltage and currents signals, indicates that voltage signals are preferable if the fault locator is installed in a large outdoor substation and the sampling frequency used by the fault locator is high.

It becomes more and more common on 132/150 kV level to use gas insulated substations when new stations are build. In this case, the internal reflections on the busbars become quite different compared internal waves on outdoor substations.

Due to the compact size of the equipment, the surge impedance of the GIS busbar becomes more comparable to the surge impedance of a high voltage cable. A surge impedance for typical gas insulated busbars is typically 75–95 Ω [5, 6]. Furthermore, the length of the busbar is much smaller compared to outdoor busbars which will influence the internal busbar waves significantly. Figure 8.6a and b shows a GIS system at the Danish substation Vester Hassing and cable termination and connection of the sheath for an ABB GIS system respectively.

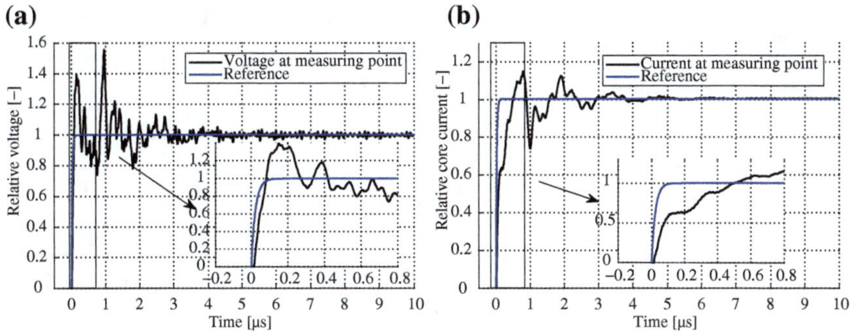

Fig. 8.7 **a** Voltage and **b** current signals simulated at the cable termination in a GIS station

It is noticed from Fig. 8.6b that the sheath is terminated in a manner that allows mounting of a current coil wherefore the sheath current signals can be used for fault location on GIS systems as well.

Similar to station Nibstrup studied earlier in this section, the Danish GIS station Karstrup has two cable lines and transformers connected. A single line diagram of the station can be found in Appendix D. The transformer is placed in a separate building and connected with three 62 m single-core coaxial cables. A simplified model of the station is implemented in PSCAD/EMTDC. The busbars, bushings, switches and breakers are modelled using the Bergeron line model with surge impedances used for typical GIS equipment found in [5, 8, 9]. The shunt capacitance of bushings, measuring transformers and switches are included based on the method presented in [6]. The cables used to connect the transformer are modelled using the frequency dependent phase model. The shunt capacitance of the transformer is included as described in the previous section. The simulation time step is set to 1 ns. A fault voltage wave is applied at the first cable termination and the transient response at that location is shown in Fig. 8.7.

Due to the short busbars, the inter-busbar reflections that occur are faster and tend to cancel faster as well. However, reflections from the transformer situated 62 m away from the GIS contributes to the voltage and current profiles as well. This makes the time it takes for the inter-busbar waves to settle prolonged. Whether or not the influence of the substation's busbars can be neglected or not is very substation dependent both for outdoor- and GIS-stations. An evaluation for each station is of interest until more experience is gained.

8.3 Fault Wave Reflection and Refraction

When developing fault location methods for crossbonded cables, it is important to know how different power system components affect the fault waves. In this section, the fault wave behaviour when encountering the most common discontinuities present

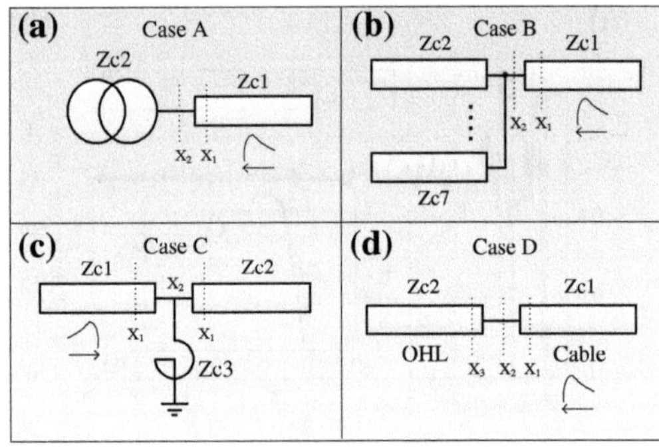

Fig. 8.8 Case study definition. **a** Between cable and transformer, **b** between one cable and *n* additional cables, **c** between two cables at shunt reactor location, and **d** between cable and OHL

in the future Danish transmission network is studied. The following four cases are selected:

- Assuming a negligible influence of the substation's busbars (small station or GIS station), the transient behaviour at the fault locator terminals is determined by the number of cables and OHLs connected to the substation.
  In the future Danish network, the number of cables connected to a substation at the same voltage level vary substantially. In Jutland, between one and four cables are typically connected to each station whereas in the area surrounding Copenhagen more than six cables can be connected in to the same substation (see Fig. 1.1). The cables in Jutland are typically longer, but fewer are connected to each station, whereas on Sealand, the cables are shorter, but more are connected to the same terminal. How faults waves are reflected and refracted at these terminals great important for fault location purposes.
- Discontinuities, other than the crossbondings, along the crossbonded cable is of interest for fault location purposes as well. These discontinuities can contribute to the characteristics of the fault waves that arrive at the fault locator terminals and their effect must be studied to identify potential problems. For connections to offshore wind farms, shunt reactors can be sometimes installed along the cable line to compensate for reactive power production of the cable. For fault location purposes, it is quite important to know how fault waves are reflected and refracted at these terminals.
- Hybrid lines will be in operation in the future Danish grid. At the interconnection point between the OHL and cable section, reflections similar to those studied in Sect. 8.2 are expected. This can affect the ability to locate the fault significantly, as only a fraction of the incoming fault wave is transmitted.

Based on the description above, four cases with common discontinuities are defined in Fig. 8.8.

Case A is quite well studied in this report already; a single cable/single power transformer configuration where the voltages and currents in the connection point is of interest. Case B involves a wave travelling on one cable terminated into a number of additional cables. In this study, six additional cables are selected. Case C studies a shunt reactor connected between two cable sections. This configuration is used several times for compensating long lines as for instance links to offshore wind farms. The last case involves the interconnection between a cable section and an OHL section.

As shown in Sect. 7.1, a fault between core and sheath causes a coaxial wave to travel on the faulted cable towards the cable ends. The behaviour of such a wave at different discontinuities can be studied by applying a surge between core and sheath at a distance from the discontinuity and observe the behaviour just before the discontinuity, at the discontinuity and in some cases after the discontinuity. The cases shown in Fig. 8.8 are implemented in PSCAD/EMTDC with use of the 165 kV case study cable (see Sect. 4.1). The transformer and reactor are modelled as described in Appendix B, but the stray capacitance is neglected. The configuration of the OHL is described in Appendix B as well. The voltage and core currents on the cable for the four cases are presented in Figs. 8.9 and 8.10 where the time scale is shifted so the wave reaches the discontinuity at $t = 0$.

8.3.1 Case A

As mentioned, this case is already well discussed. Core voltage signals are the only option for accurate fault location as no current wave can penetrate the transformer windings instantaneously. Figures 8.9a and 8.10a shows a doubling of the incoming voltage wave and a current wave that is reduced to zero.

8.3.2 Case B

Case B study the reflection between different numbers and types of transmission lines. For a lossless line, the asymptotic surge impedance between core and sheath of a single coaxial cable is calculated as shown in Eq. 8.1.

$$Z_{0C} = \sqrt{\frac{\mu_0}{\epsilon_0 \epsilon_r \pi^2}} \ln\left(\frac{r_2}{r_1}\right) \tag{8.1}$$

where r_1 is the inner radius of the core, and r_2 is the outer radius of the sheath. The ratio r_2/r_1 varies for different high voltage cables; values for the surge impedance between 20 and 60 Ω are found in the literature.

The surge impedance for the 165 kV case study cable is 24.8 Ω. With the six additional cables connected to the station, Figs. 8.9b and 8.10b show a voltage wave of 0.28 pu and a current wave of 1.71 pu. Figure 8.11a and b shows the resulting voltage and current at the discontinuity for an incoming wave of 1 pu as function of the number of additional equal cables matching with the results obtained in the simulations.

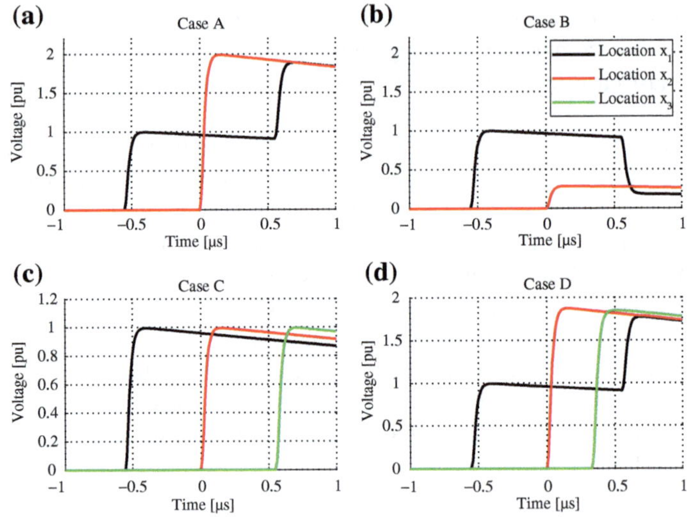

Fig. 8.9 Voltage waves at location x_1, x_2 and x_3 for the case studies defined in Fig. 8.8

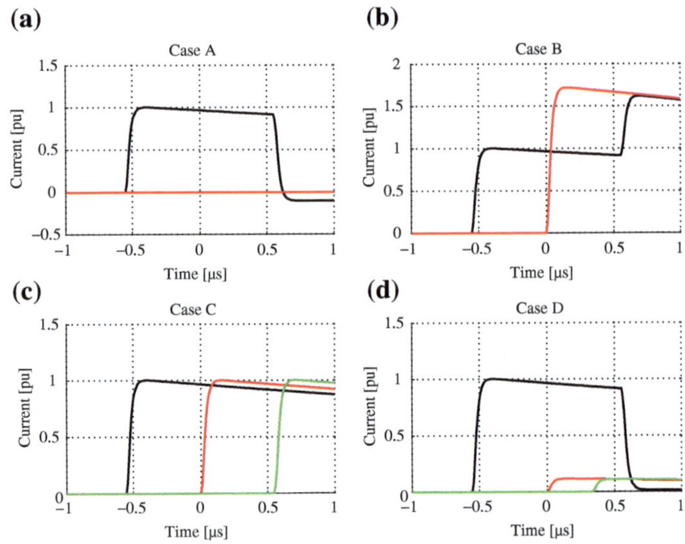

Fig. 8.10 Current waves at location x_1, x_2 and x_3 for the case studies defined in Fig. 8.8

As the number of cables increases, it becomes more beneficial to use either core or sheath current signals as these are reflected with a reflection coefficient approaching 1 resulting in a doubling of the incident current wave.

OHLs have very different surge impedances compared to cables. As the value is larger, more OHLs are needed for having low equivalent station surge impedance.

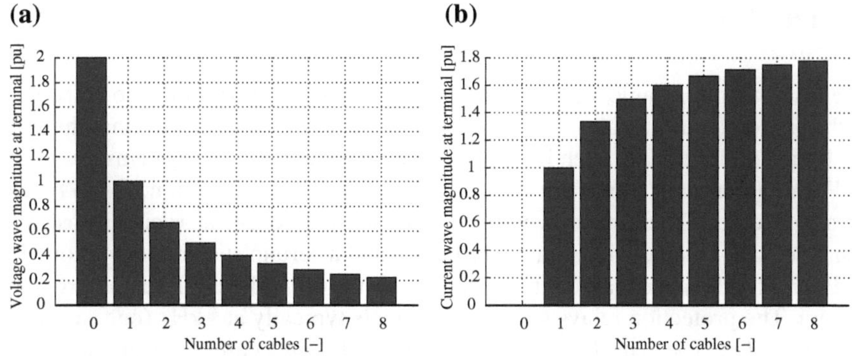

Fig. 8.11 Resulting **a** voltage and **b** current at the discontinuity for an incoming wave of 1 pu as function of the number of additional equal cables

A rough estimate of the substations combined surge impedance can be determined in each case or it can be chosen to record both all three voltages and either the three core or sheath currents.

8.3.3 Case C

Assuming a pure inductive reactor, the reactor windings will act as an open end at the instance the high frequency wave arrives. An incoming wave will meet the combined surge impedances of the shunt reactor and the additional line and the combined surge impedance is therefore determined by the surge impedance of the additional lines only at high frequencies. Both voltage and current waves will continue unaffected making the reactor station well suited for an additional measuring point effectively reducing the length of the monitored line (benefits shown in Sect. 8.1). This is shown in Figs. 8.9c and 8.10c

8.3.4 Case D

The interconnection of a cable and an OHL gives rise to a large change in surge impedance and thus a large reflection. A voltage wave coming from the cable section into the cable is increased in magnitude with a factor of approximately 1.85. The current is reduced to 0.15. A wave traveling the opposite directing will be reflected with opposite sign. This can be a problem on hybrid lines, as fault created waves are strongly affected at the discontinuity and only a fraction of the expected wave will arrive. This is examined in a later section where fault location on hybrid lines is examined.

8.4 Fault Inception Angle

The fault inception angle for faults in an OHL-system can vary from 0 to 360 degrees as the cause of fault in many cases are external factors. Due to the fact that the cables are buried underground, such incidents do not happen for cable systems.

For an internal or joint fault to occur in a cable, the insulation needs to be electrically stressed. The breakdown will thus most likely occur near either positive or negative peak voltage with some distribution and a certain standard deviation [10].

Energinet.dk records all fault transients with the use of line differential protection system. The protection relays sample the signals typically at 1 kHz (one sample per 18° at 50 Hz). The data from three 165 kV cable fault cases is examined, and it is found that for all cases, the fault occurs with a maximum deviation from peak voltage of 25°. A fault occurring at peak voltage will give rise to the largest fault transient, but as the fault inception angle deviates from 90°, the magnitude of the fault voltage wave is lowered and the waves can be harder to pick up at the FLT. The magnitude of the core and sheath current waves are for internal faults determined by the ratio between the instantaneous voltage at the moment of fault and the coaxial surge impedance.

As shown in Sect. 8.3, voltage waves are best used if a single line/single transformer configuration is used. If many additional cables are installed, either core or sheath currents are preferable. In the case where a single cable with the same surge impedance is installed, the voltage and current waves are equally beneficial. This is, in case of a fault inception angle different from 90° on a long cable, the worst case as the incoming wave is complectly transmitted. As mentioned, it is expected that an internal fault will always occur near peak voltage. However, some deviation can occur and the influence needs to be tested. Faults with inception angles of 25°, 30° and 45° are simulated on a 60 km long cable with the fault location 1 km from Terminal A. The core voltage and core currents at the two cable ends as well as the waves at the fault location are shown in Fig. 8.12. The sheath currents are not shown as they are equal to the core currents as long as only coaxial waves contribute to the current profile.

The voltage fault wave at the fault location for the three inception angles of 25°, 30° and 45° are 93.64, 66.15 and 55.91 kV. The coaxial surge impedance determines the magnitude of the initial fault current wave on the core and is calculated by Eq. 8.1 as 24.8 Ω. The resulting core and therefore also sheath currents waves are 3.77, 2.67 and 2.25 kA.

The frequency content in the fault signals at the FLTs is not changed as function of fault inception angle and is equal in all signals. However, in noisy signals, the superimposed fault waves can be easier to distinguish from the power frequency signals if the magnitude of the fault waves is larger. The fault location is in this particular case for all fault inception angles determined with an error of 8–10 m.

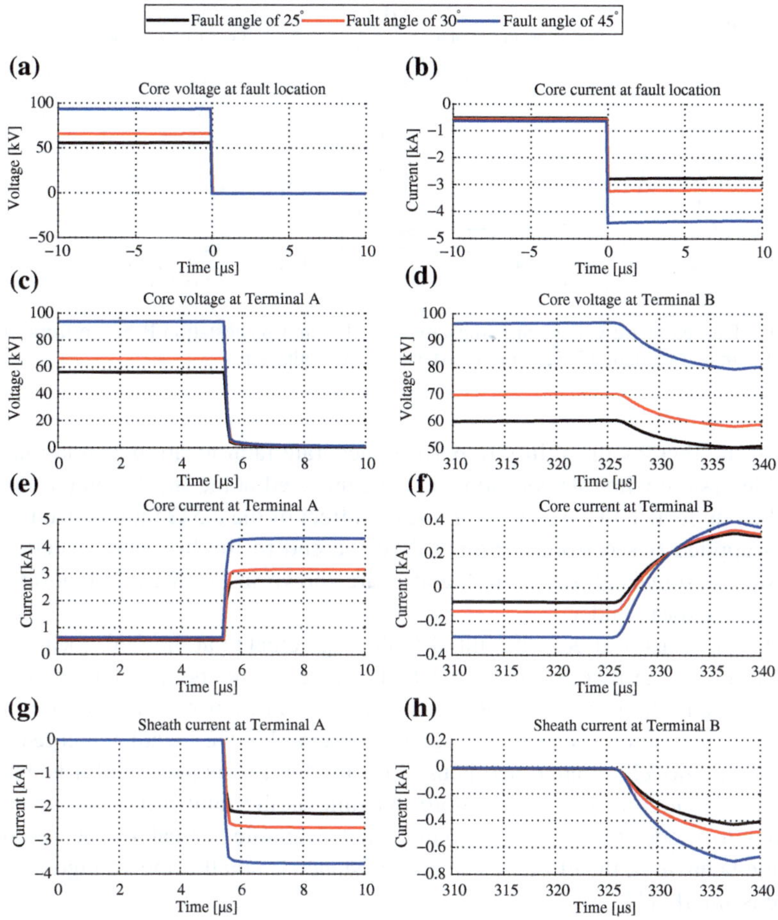

Fig. 8.12 Faulted phase **a** core voltages and **b** core currents at the fault location, **c** core voltages at Terminal A, **d** core voltages at Terminal B, **e** core currents at Terminal A, **f** core currents at Terminal B, **g** sheath currents at Terminal A and **h** sheath currents at Terminal B for a fault inception angle of 25°, 30° and 45°

8.5 Fault Arc Resistance

It is general practice to neglect the arc voltage drop in cables for travelling wave-based fault location studies as the arc is very short with resulting neglectable arc impedance [6]. However, some references argue that the pressure can build up causing the fault arc voltage drop to increase [11, 12]. It is very difficult to predict how the dynamics of such an arc will behave due to the complex mechanisms involved. In a simplified first approach, a constant resistance is added at the fault location ($R_f = 2.5\,\Omega$). Because the initial fault wave is now different from a fault wave created with zero fault resistance, it is of interest to examine whether this will affect the first wave at

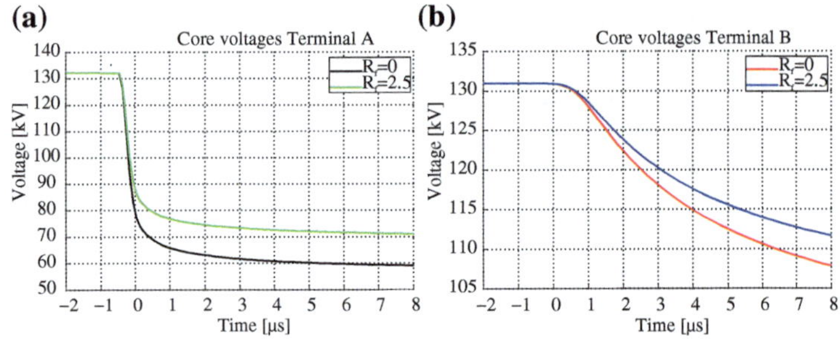

Fig. 8.13 Faulted phase voltage at **a** substation A 2 km and **b** substation B 58 km from the fault location with and without a 2.5 Ω fault resistance in the fault loop

the FLT and thereby the ability to perform accurate fault location estimation. A 60 km cable system with one additional cable connected at the substation (worst case) described in Sect. 8.3 is used to examine the effect of the 2.5 Ω fault resistance. The fault is applied 6 km from Terminal A at the end of the first major section, and the resulting signals at Terminal A and B are shown in Fig. 8.13 with and without the fault resistor included.

The fault resistor affects the initial fault surge created at the fault location, thus affecting the surge that arrives at the FLT. However, for both signals recorded at Terminal A and B, the ability to determine the arrival instance is only slightly if at all affected by the fault resistance. The same is observed in the sheath and core current.

A simple modelling approach is used for the fault resistance, and it is expected that the behaviour will be much more dynamic as the fault progresses over time. Methods to model this are unfortunately not available for cables, but even with dynamic behaviour, the effect of the arc resistance is small if the voltage drop over the arc is small [11].

8.6 Sensitivity of the Coaxial Modal Wave on Cable and Cable System Parameters

It is important to determine which cable and cable system parameters affect the coaxial modal velocity and attenuation as this modal signal seems preferable for fault location purposes. It is unbeneficial if the velocity of the chosen mode will change over time and change along the cable route due to changes in, cable, or cable system parameters. Furthermore, damping in real-life cables, which is not accounted for in the model, is a potential problem when the development of a fault locator system is based on simulations. In this section these issues are addressed.

8.6.1 Coaxial Modal Wave Velocity

Because the magnetic field is assumed perfectly confined within the metal sheath at high frequencies, the coaxial wave velocity is not dependent on layers or materials outside this layer. The conductivity of a good conductor like copper and aluminium is very high. As a result, the skin depth will be very low for these materials at high frequencies (see Eq. 6.15). The electromagnetic waves will consequently decay very rapidly into the material, and the current is almost purely flowing on the surface of the conductor. As a result, the coaxial wave velocity becomes almost independent of the core and sheath materials.

As described, the magnetic and electric fields are confined in the homogenous dielectric. Due to this, the coaxial wave velocity is highly dependent on the characteristics of this material [13]. Assuming a perfect dielectric (no conductivity), the imaginary part of the dielectric constant $\epsilon(\omega) = \epsilon^{'}(\omega) + j\epsilon^{''}(\omega)$ is zero. Furthermore, if $\epsilon(\omega)$ is considered frequency independent in the frequency range of interest, the coaxial wave velocity can then be calculated at high frequencies by use of Eq. 8.2 without including the complex calculations involved in the modal decomposition [14].

$$v_c = \frac{c}{\sqrt{\mu_{r,i1}\epsilon_{r,i1}}} \qquad [\mathrm{m/s}] \tag{8.2}$$

where c is the speed of light, $\epsilon_{r,i1}$ ($\epsilon^{'} = \epsilon_0\,\epsilon_{r,i1}$) is the relative permittivity of the main insulation material and $\mu_{r,i1}$ is the relative permeability ($\mu = \mu_0\,\mu_{r,i1}$). The relative permeability $\mu_{r,i1}$ is 1 for non-magnetic materials as used in the cable and the permittivity $\epsilon_{r,i1}$ is 2.5 for pure XLPE [15].

As shown in for instance Fig. 4.1, two semi-conductive layers surround the main insulation. It is assumed that the semi-conducting layers carry only a negligible longitudinal current as their conductivity is required by norm to be higher than 1 mS/m [3]—values between 0.01 and 10 Ωm are, however, more common [16]. In [14], it is shown that the coaxial wave velocity is not a function of the conductivity of the semi-conducting layers in the frequency range of interest and in [17], it is concluded that the effect is very small for small semi-conducting layer thickness and high resistivity. Therefore, variations in this material parameter is not of special interest for the coaxial velocity for normal high voltage cables. As all studies in the mentioned references are based on analytical calculations of very complicated physical phenomena, field measurements are conducted later in this thesis in order to verify the claims (Sect. 10.4).

In most simulation programs, the semi conducting layers are not included. The insulation seems therefore thicker than its actual value and the capacitance becomes smaller. The effect of the semi-conducting layers are included by adjusting the permittivity of the main insulation layer and assuming the model seen in Fig. 8.14a.

The adjustment of the relative permittivity is done under the assumption that an equivalent capacitor which value is calculated using only the permittivity of the main insulation will be accurately representing the series coupling of the three capacitors

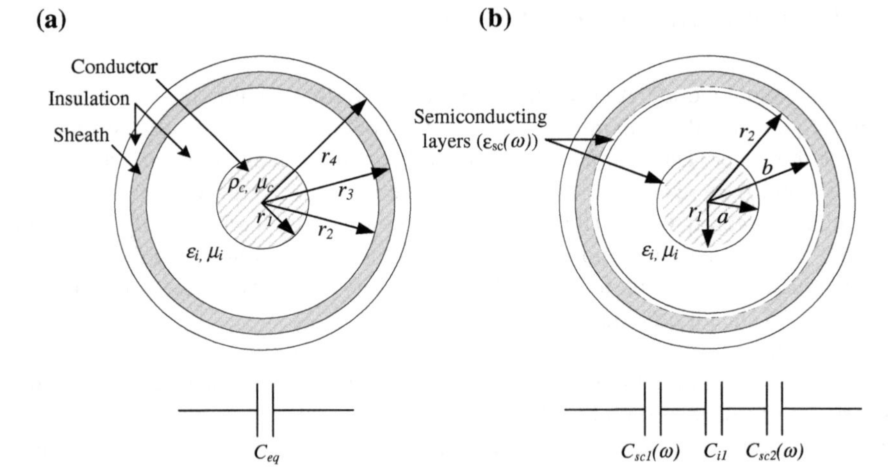

Fig. 8.14 **a** Cable model without semiconducting layers, and **b** cable model including the semiconducting layers

shown in Fig. 8.14b. Equation 8.3 shows the calculation method for the equivalent capacitance that represents the main insulation and semiconducting layers [3] assuming a perfect dielectric.

$$C_{eq} = \frac{2\pi\varepsilon_0\varepsilon^*}{\ln\left(\frac{r_2}{r_1}\right)} \qquad [\mathrm{F/m}] \tag{8.3}$$

where ϵ^* is calculated as:

$$\epsilon^* = \epsilon_{r,i1}\frac{\ln(\frac{r_2}{r_1})}{\ln\left(\frac{b}{a}\right)} \qquad [-] \tag{8.4}$$

In Eq. 8.4, $\epsilon_{r,i1}$ is the relative permittivity of the insulation material (XLPE), and r_1, r_2, a and b are defined in Fig. 8.14b.

Frequency Dependent Permittivity

Because XLPE is a non-polar material, its permittivity is considered almost frequency independent in the frequency range of interest [18]. The permittivity of the semiconducting layers is, however, because of the carbon content, strongly dependent on frequency, and an equivalent permittivity representing the three layers should therefore be considered frequency dependent as well. This is not taken into account in Eq. 8.2 or any of the current simulation cable models, and the effect is studied. The relationship between permittivity and frequency for a typical semi-conducting

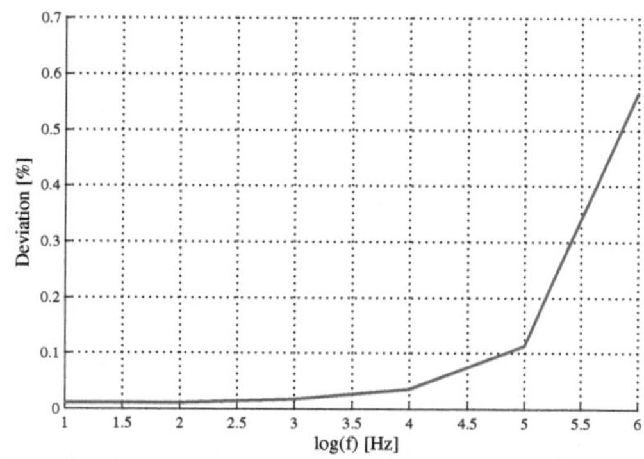

Fig. 8.15 Deviation between the equivalent capacitance calculated with an adjusted permittivity and one including the effect of the the frequency dependent semi-conducting layers as function of frequency

material is discussed in [19]. The authors show that the permittivity of the semi-conducting layers is quite constant up to 10 kHz with a value around 1000 whereafter it drops, and at 1 MHz, it is approximately 100. Assuming only displacement current in the dielectric, the capacitance between the conductor and the sheath including the frequency dependent semiconducting layers is calculated as shown in Eq. 8.5 [20].

$$C_{eq}(\omega) = \frac{2\pi\, \varepsilon_{sc1}(\omega)\, \varepsilon_{r,i1}(\omega)\, \varepsilon_{sc2}(\omega)}{\varepsilon_i\, \varepsilon_{sc2}(\omega) \ln\left(\frac{a}{r_1}\right) + \varepsilon_{sc1}(\omega)\, \varepsilon_{sc2}(\omega) \ln\left(\frac{b}{a}\right) + \varepsilon_{sc1}(\omega)\, \varepsilon_i \ln\left(\frac{r_2}{b}\right)} \tag{8.5}$$

where $\varepsilon_{sc1}(\omega)$ and $\varepsilon_{sc2}(\omega)$ are the frequency dependent permittivities of semi-conducting layers, and the physical parameters used are defined in Fig. 8.14b.

Whether it can be correctly assumed that Eqs. 8.3 and 8.5 give equal results depends on whether the capacitances of the semi-conducting layers are many times larger than the capacitance of the insulation material. For a high capacitance of the semi-conducting layers, the resulting admittance will tend to act as a short, and the total core/sheath admittance will be dominated by the insulation admittance [17]. In Eq. 8.5, this means that the term $\varepsilon_{sc1}(\omega)\, \varepsilon_{sc2}(\omega) \ln\left(\frac{b}{a}\right)$ is many times larger than the two others. As seen, this is true when the frequency dependent permittivity of the semi-conducting material is larger than the permittivity of the insulation and the insulation layer is thicker than the semi-conducting layers. The latter is always the case, but the former assumption can be violated as the frequency dependent permittivity of the semi-conducting layers drops as the frequency increases. This can lead to a violation of the initial assumption for using Eq. 8.3. To examine if this is a problem at high frequencies, the deviation between the resulting capacitance calculated using the two methods is determined, and the result is shown in Fig. 8.15.

Figure 8.15 shows that the capacitance of the equivalent insulating material (C_{eq}) will represent the equivalent layer, comprising of the primary insulation material and the two semiconducting layers, well in the frequency range of interest. At 1 MHz ($\varepsilon' = 100$), the error is 0.36 % (plus indicated that the capacitance calculated the adjusted method is the largest) and 0.02 % at 50 Hz ($\varepsilon' = 1000$). As the deviations are small, the adjustment method can be used for both simple and modal calculations

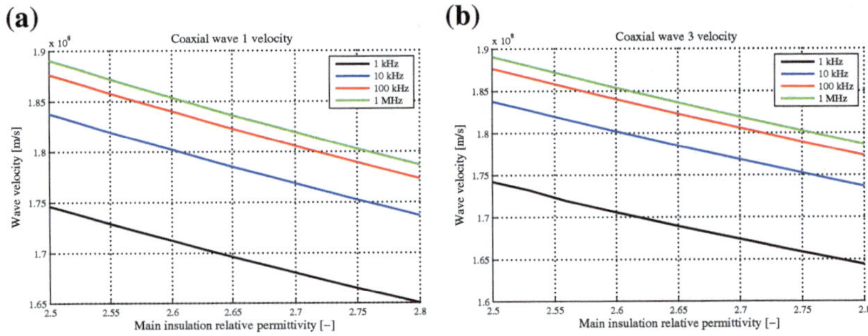

Fig. 8.16 **a** Coaxial mode 1 and **b** Coaxial mode 3 wave velocity as function of main insulation relative permittivity evaluated at different frequencies

without introducing much error. However, as input for a travelling wave fault locator system, a coaxial velocity which is measured on the real-life cable should be given as input. The result is confirmed by field measurements in [21] where it is shown valid up till 1 MHz.

Sensitivity of Coaxial Modal Velocity on Relative Permittivity

The data sheet gives only minimum values or typical values of the cable layers' dimensions, and, as the corrected relative permittivity, and therefore the coaxial wave velocity, is dependent on deviations on these parameters, a sensitivity analysis is carried out. The coaxial mode 1 and 2 wave velocities as function of main insulation relative permittivity evaluated at different frequencies are shown in Fig. 8.16a and b.

The coaxial modal velocity's dependence on a parameter which is corrected using several assumptions indicates that field measurements are preferable in order to determine the correct coaxial wave velocity. During this PhD, several methods were developed for measuring the coaxial wave velocity. These are described in Sect. 10.4 where field measurement are presented.

Temperature Dependent Permittivity

The permittivity of cross-linked polyethylene (XLPE) is temperature dependent. However, in the temperature range where cables are normally operated, the dependence is minor and needs not to be considered [18, 22]. The semi-conducting material is temperature dependent as well, but as shown in the previous section, the equivalent capacitance between core and sheath is dominated by the capacitance of the main insulation which makes the effect less important.

8.6.2 Attenuation of the Coaxial Modal Wave

The attenuation of the coaxial modal signal is dependent only on material and layers inside the sheath layer. The material of the core and sheath will have an affect, but can be well predicted by the current cable models. The attenuation introduced by the semi-conducting layers is not represented in the model implemented in PSCAD/EMTDC, but a fixed shunt conductance can be given as an input. How the semi-conducting layers affect the coaxial damping is discussed in [3, 23, 14, 24]. The authors propose a model of a conductance in parallel with the capacitance representing the capacitance of the semi-conducting layers and conclude that the attenuation effect is dependent on the carbon content. The author of [3] concludes that "With the high permittivity (1000), the capacitance tends to short out the conductance, and no appreciable increase of the attenuation is seen". The same can in general be concluded for a permittivity of 100 which is the minimum value for the frequency range of interest. The statement is examined in a later section where field measurements of pulse propagation on crossbonded cables are presented.

8.7 Determination of the Modal Velocities

The coaxial wave velocity is relatively simple to measure in real life. As seen in Fig. 6.4b, the three unique coaxial mode signals travel with the same velocity at frequencies above approximately 1 kHz and become frequency independent above 10 kHz. Furthermore, as this type of wave is the fastest at any frequency, the first signal change noticed at any location will be due to a coaxial wave.

Several different methods to measure the coaxial wave velocity are developed and described in Sect. 10.4. Amongst them are both single ended and a double ended synchronised methods. Field tests are carried out and the velocity extracted with good precision. These field tests are described in detail in Sect. 10.4. As the field is confined between the core and sheath, the measurements can be carried out before and after installation.

Compared to the coaxial waves, the intersheath mode waves are more problematic to determine. First, as the wave velocity is dependent on several parameters, it can change as function of location. Depending on the measuring method used, either a velocity for one to a few minor sections is used or a mean value for the whole cable route. Both can locally be non-representative for the true velocity, and thus, errors are introduced if the signals are used.

8.8 Measuring Transformers

Both core voltages, core currents and sheath currents are of interest for fault location purposes. However, it is important how the measuring transformers used today represent the high frequency fault transients in the high voltage signals in the secondary

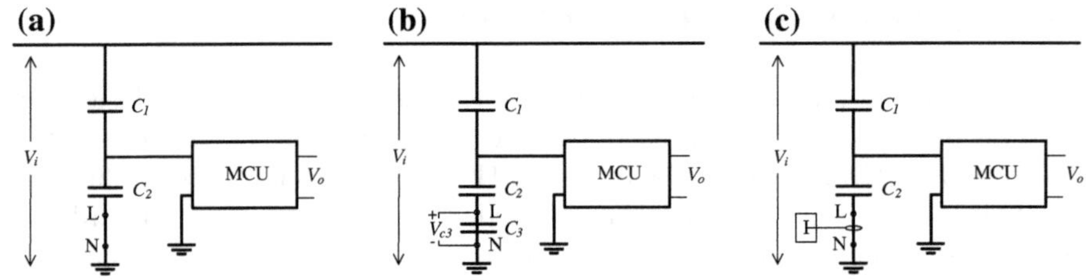

Fig. 8.17 **a** Capacitive voltage instrument transformer, **b** transients voltage measurements by additional capacitor and **c** transients voltage measurements by current transducer

circuit. Because the two-terminal method is recommended, only the arrival of the first fault wave needs to be detected. It is not important if the fault transient is represented correctly in the secondary circuit. The only requirement is that the wave is not delayed in a degree that compromises fault location and not damped to a degree that it becomes undetectable. These issues have been discussed by several authors and a quick review is given here.

8.8.1 Capacitive Voltage Transformers

Capacitive voltage transformers are known to have low bandwidth and are especially designed for representing only the power frequency in the secondary circuit. The capacitive voltage transformer consists of two capacitors in series with a magnetic coupling unit (MCU) connected between the two. The principle diagram is shown in Fig. 8.17a.

The magnetic coupling unit is the cause of the poor frequency response of the capacitive voltage transformer [25]. However, it is possible to access the neutral of the voltage divider and either connect an additional capacitor between terminals L and N and measure the voltage across it or apply a short and measure the current through the divider (Fig. 8.17b and c).

Practical experience with adding an additional capacitor in series with the voltage divider is reported by researchers in Romania in the frame of a project called "Transient disturbances in Romanian open-air high voltage substations". Good results are obtained and the method is recommended for transient voltage measurements [26].

The current flowing through the voltage divider is determined by the total capacitance of the divider and the time derivative of the input voltage. In case of a high frequency transient in the high voltage signal, a high frequency current will flow through the capacitors. This current can be measured and used as input to the fault locator unit as shown in Fig. 8.17c. If a representation of the high voltage signal is desired, the divider current can be integrated. However, as only the first wave at the fault locator terminal is needed, applying integration will only lower the bandwidth and better results are obtainable if no integrator is used.

The use of current measurements in capacitive voltage dividers is proposed for measurements of voltage for harmonic analysis [27, 28].

8.8.2 Inductive Voltage Transformers

The use of the classical inductive voltage transformer is not very well studied for travelling wave-based fault location purposes. The use of an inductive coupler can compromise the accuracy of the fault location due to the band-pass features of the voltage transformer [29, 30]. In [31], the transformer is used for transients up till 10 kHz. Measurements are provided showing good transient performance. From the literature published in the field, a conclusion on the performance on the inductive voltage transformer is difficult to make; more experience is needed. This is discussed in the section summary.

8.8.3 Inductive Current Transformers

The use of the inductive current transformers for transient representation in the secondary circuit is quite well studied. Leakage inductance and stray capacitance as well as saturation of the ferromagnetic core limit the bandwidth of the transformer. Nonetheless, good representation of high frequency components of the high voltage signals in the secondary circuit is reported in [32–35]. It is concluded that the transformers can be used up till 200 kHz which according to the authors is suitable for travelling wave-based fault location.

8.8.4 Rogowski Coils

Low voltage high bandwidth Rogowski coils can be used to measure the sheath current for systems with one or more additional lines connected at the same substation as the monitored cable.

A Rogowski coil consists of a helical coil wound on a non-magnetic core with the core encapsulating the current it measures. When a current flows through the coil, a voltage proportional to the rate of change of current is induced:

$$v(t) = -\frac{d\varphi}{dt} = \mu_0 n A \frac{di}{dt} \tag{8.6}$$

where n is turns per meter, A is the cross-sectional area of the coil and i is the encapsulated current.

By adding a wire loop wound in the opposite directing to the first loop, the Rogowski coil becomes immune to the influence of nearby current carrying conductors. This is traditionally done in most commercial available Rogowski coils.

By integrating the voltage induced by the coil, a measure of the current through the coil is obtained. However, as for the capacitive voltage divider, only the first fault transient needs to be detected for fault location purposes. Therefore, by removing the integrator, a direct measure of the rate of change of current can be obtained and thus a higher bandwidth. Furthermore, by removing the integrator, only the passive coil is left and no power supply is necessary.

The bandwidth of Rogowski coils is limited by the series inductance of the wire and by stray capacitance [36, 37]. The propagation time delay of the coil increases with frequency [38]. For some Rogowski Coil designed especially for high frequency purposes, the frequency response can be much higher than 1 MHz [39]—up to 30 MHz is not uncommon. This makes them ideal for travelling wave-based fault location. Furthermore, they are easy to mount on the cable's termination as already shown in Fig. 7.10, and, relatively cheap.

8.8.5 Summary

Sheath currents can be measured using Rogowski coils and core voltage signals can with benefit be measured with use of capacitive voltage transformers. The transient response of both transducer types make them well suited for use with a travelling wave-based fault location system. The two transducer types cover in principle all system configurations as currents are beneficial to use if two or more additional lines are installed at the substation and core voltages are beneficial as input for single transformer/single line configurations. However, due to practical and economic issues, capacitive voltage transformers are not always installed. On the 132/150 kV level, it is common to use inductive voltage transformers due to their price and ability to discharge the cables after the are disconnected from the rest of the grid. The transient response of inductive voltage transformers is not well studied for high frequency transients. Due to the lack of research in the area, it is recommended to provide a reference signal during the test period of a potential fault locator either in the form of a sheath current measured with a Rogowski coil or a voltage measured with a capacitive voltage transformer.

The literature promotes the inductive current transformer as capable of representing the high frequency transients with good accuracy in the secondary—signals up till 200 kHz can be replicated. Compared to for instance the Rogowski coil, this bandwidth is low, and how it will affect the accuracy of a fault location system for crossbonded cables is not known exactly. Therefore, it is again recommended to provide a reference signal under test until the performance of the inductive current transformer can be fully evaluated.

The travelling waves are affected by secondary circuit parameters as cable type and length, the transformer parameters, and the resistance and inductance of the relay burden [40, 41]. Therefore, dedicated travelling wave transducers as the Rogowski coil or a capacitive voltage divider with an output merely for travelling wave measurements are preferable, as no interference from the protection or measuring system is introduced.

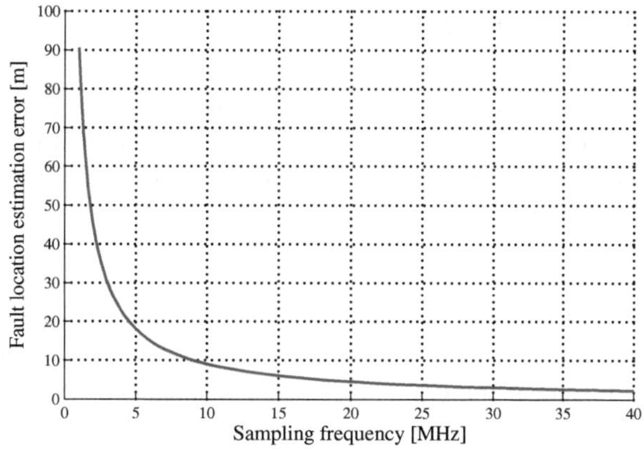

Fig. 8.18 Fault location estimation error for the two-terminal synchronised fault location as function of sampling frequency

Based on this literature study, it is recommended to use either Rogowski coils for measuring sheath currents or to use capacitive voltage dividers to measure core voltages.

8.9 Fault Locator Sampling Frequency

The sampling frequency of the data acquisition units installed will influence the accuracy of the two-terminal fault location method. If $\Delta\tau_A$ and $\Delta\tau_B$ represents the delay between the exact wave arrival instance at the terminals and the instance of sampling at Terminal A and B, the estimated error introduced by the sampling delay alone can be calculated as shown in Eq. 8.7.

$$x_e = \frac{v_m \cdot (\Delta\tau_A - \Delta\tau_B)}{2} \quad [\text{m}] \tag{8.7}$$

Equation 8.7 shows that if the delay is equal at both terminals, the correctly fault location is determined. The worst case occurs when the arrival instance is determined correct at one terminal ($\Delta\tau = 0$) and the other is one sample delayed ($\Delta\tau = 1/f_s$). The theoretical worst case fault location estimation error as a function of sampling frequency is plotted in Fig. 8.18.

As expected, the theoretically largest error can occur at a low sampling frequency. The estimation error at 1 MHz is 86 m while an error at 40 MHz is 2.1 m. The choice of sampling rate is an economical issue, and it is determined by the operator of the transmission system.

8.10 Summary

In this section, the parameters influencing the accuracy two-terminal fault location method are discussed. It is found that the method is unaffected by most parameters as fault inception angle and fault resistance. Damping of the high frequency fault components increases as the length of the cable line increases. This is a potential problem for long lines, but for most lines planned in the future Danish grid, this is not an issue.

For large substations, the influences of the busbars on the incoming fault wave should be evaluated when the choice of input signal is made. For GIS stations, the influence is minor and the choice will depend on the number of lines connected the substation.

The coaxial wave is after an analysis still the preferred mode to use for fault location purposes as the parameters that affects the velocity are constant or are only changing slightly in the working area.

The sampling frequency determines the horizontal resolution of the fault locator units and is therefore an important parameter for the accuracy of the fault location. The choice is, however, economical as very fast commercially sampling equipment is available on the market.

References

1. C.S. Indulkar, P. Kumar, D.P. Kothari, Sensitivity analysis of modal quantities for underground cables. Gener. Transm. Distrib. IEE Proc. C **128**(4), 229–234 (1981)
2. C.S. Indulkar, P. Kumar, D.P. Kothari, Modal propagation and sensitivity of modal quantities in cross-bonded cables. Gener. Transm. Distrib. IEE Proc. C **130**(6), 278–284 (1983)
3. B. Gustavsen, Panel session on data for modeling system transients insulated cables. in *Power Engineering Society Winter Meeting, 2001. IEEE*, vol. 2 (2001), pp. 718–723
4. U.S. Gudmundsdottir, PhD-thesis, *Modeling of Long High Voltage AC cables in Transmission Systems*, 2nd edn. (Fredericia, Denmark, Energinet.dk, 2010) ISBN 978-87-90707-73-6
5. S. Matsumura, T. Nitta, Surge propagation in gas insulated substation. IEEE Trans. Power Apparatus Syst. **PAS-100**(6), 3047–3054 (1981)
6. D. Povh, H. Schmitt, O. Valcker, R. Wutzmann, Modelling and analysis guidelines for very fast transients. IEEE Trans. Power Deliv. **11**(4), 2028–2035 (1996)
7. ABB, *Product Brochure—Gas-insulated Switchgear ELK-04 Modular System up to 170 kV, 4000 A, 63 kA* (Hanau-Großauheim, Germany, 2012)
8. S.y. Jung, J.y. Kim, C.s Oh, B.-W. Lee, J.-Y. Koo, Measurement of propagation velocity of the electromagnetic wave produced by partial discharge in a gas-insulated switchgear. in *Telecommunications Energy Conference, 2009. INTELEC 2009. 31st, International* (2009), pp. 1–7
9. S. Okabe, S. Kaneko, M. Yoshimura, H. Muto, C. Nishida, M. Kamei, Propagation characteristics of electromagnetic waves in three-phase-type tank from viewpoint of partial discharge diagnosis on gas insulated switchgear. IEEE Trans. Dielect. Electr. Insul. **16**(1), 199–205 (2009)
10. X. Zhihan, *Fault location and incipient fault detection in distribution calbes* (The University of Western Ontario London, Ontario, Canada, 2011)
11. G. Ziegler, *Numerical Distance Protection—Principles and Applications*, 3rd edn. (Publicis Kommunikation Agentur GmbH, GWA, Erlangen, 2008)

12. C.k. Jung, J.b. Lee, Fault location algorithm on underground power cable systems using noise cancellation technique. in *Transmission and Distribution Conference and Exposition, 2008. IEEE/PES* (2008), pp. 1–7
13. H.L. Heck, S.H. Hall, *High Speed Digital Designs*, (Wiley, Hoboken, 2009)
14. W.L. Weeks, Y.M. Diao, Wave propagation characteristics in underground power cable. IEEE Trans Power Apparatus Syst. **PAS-103**(10), 2816–2826 (1984)
15. IEC, *IEC 60287 3 2 Electric Cables Calculation of the Current Rating Part 3 2*, 1st edn. (IEC, Geneva, Switzerland, 1995–06)
16. Akihiro Ametani, Yukata Miyamoto, Naoto Nagaoka, Semiconducting layer impedance and its effect on cable wave-propergation and transoents characteristics. IEEE Trans. power Deliv. **19**, 1523–1531 (2004)
17. A. Ametani, Y. Miyamoto, N. Nagaoka, Semiconducting layer impedance and its effect on cable wave-propagation and transient characteristics. IEEE Trans. power Deliv. **19**(4), 1523–1531 (2004)
18. R. Liepins, C.C Ku, *Electrical Properties of Polymers: Chemical Principles*, 1st edn. (Hanser, Munich, 1987)
19. K.-M. Jager, L. Lindbom, The continuing evolution of semiconductor materials for power cable applications. Electr. Insul. Mag. IEEE **21**(1), 20–34 (2005)
20. C.R. Paul, *Analysis of Multiconductor Transmission Lines* (IEEE Press, Hoboken, 2008)
21. K. Steinbrich, Influence of semiconducting layers on the attenuation behaviour of single-core power cables. Gener. Transm. Distrib. IEE Proc. **152**(2), 271–276 (2005)
22. V. Dubickas, *Development of On-line Diagnostic Methods for Medium Voltage XLPE Power Cables*, 1st edn. (KTH School of Electrical Engineering, Stockholm, Sweden, 2009)
23. N. Oussalah, Y. Zebboudj, S.A. Boggs, Partial discharge pulse propagation in shielded power cable and implications for detection sensitivity. Electr. Insul. Mag. IEEE **23**(6), 5–10 (2007)
24. Y.H. Md Thayoob, A.M. Ariffin, S. Sulaiman, Analysis of high frequency wave propagation characteristics in medium voltage xlpe cable model. in *Computer Applications and Industrial Electronics (ICCAIE), 2010 International Conference on* (2010), pp. 665–670
25. D. Hou, J. Roberts, Capacitive voltage transformer: transient overreach concerns and solutions for distance relaying, in *canadian Conference on Electrical and Computer Engineering*, 1996, vol. 1, pp. 119-125, 26–29 May 1996
26. M. Costea, A. Marinescu, S. Coatu, D. Rucinschi, *Characterization of the Capacitive Voltage Transformers as Coupling Paths and Measuring Devices for the High-Voltage Transients in a Substation* (University Politehnica, Bucharest, Romania)
27. ABB, *Instrument Transformers Application Guide* (ABB, Ludvika, Sweden, 2013)
28. F. Ghassemi, P. Gale, T. Cumming, C. Coutts, Harmonic voltage measurements using cvts. IEEE Trans. Power Deliv. **20**(1), 443–449 (2005)
29. D. Spoor, J. Zhu, Transfer function analysis of coupling equipment used for traveling wave fault location. in *Proceedings of the International Conference Electrical Machines and Systems* (Jesu Island, Korea, 2004)
30. M.I. Samesima, J.C. De Oliveira, E.M. Dias, Frequency response analysis and modeling of measurement transformers under distorted current and voltage supply. IEEE Trans. Power Deliv. **6**(4), 1762–1768 (1991)
31. P. Imris, M. Lehtonen, Modelling of a voltage transformer for transients. in *Power Tech, 2005 IEEE Russia* (2005), pp. 1–5
32. A.M. Elhaffar, Power transmission line fault location based on current traveling waves. Doctoral Dissertation, Helsinki University of Technology Faculty of Electronics, Communications and Automation Department of Electrical Engineering, 2008
33. D.A. Douglass, Current transformer accuracy with asymmetric and high frequency fault currents. IEEE Trans. Power Apparatus Syst. **PAS-100**(3), 1006–1012 (1981)
34. A.P.S. Meliopoulos, F. Zhang, S. Zelingher, G. Stillman, G.J. Cokkinides, L. Coffeen, Jr. Burnett, R.O., J. McBride, Transmission level instrument transformers and transient event recorders characterization for harmonic measurements. IEEE Trans. Power Deliv. **8**(3), 1507–1517 (1993)

35. D.J. Spoor, J. Zhu, P. Nichols, Filtering effects of substation secondary circuits on power system traveling wave transients. in *Electrical Machines and Systems, 2005. ICEMS 2005. Proceedings of the Eighth International Conference on*, vol. 3 (2005), pp. 2360–2365
36. W. Stygar, G. Gerdin, High frequency rogowski coil response characteristics. IEEE Trans. Plasma Sci. **10**(1), 40–44 (1982)
37. Y. Cheng, Y. Yan, C Li, The high frequency characteristic of wideband rogowski coil with asymmetric windings. in *Electrical Insulation and Dielectric Phenomena (CEIDP), 2010 Annual Report Conference on* (2010), pp. 1–4
38. W.F. Ray, C.R. Hewson, J. M. Metcalfe, High frequency effects in current measurement using rogowski coils. in *Power Electronics and Applications, 2005 European Conference on* (2005), p. 9
39. IEEE PSEC Special Report, *Practical Aspects of Rogowski Coil Applications to Relaying* (IEEE, 2010). Available at http://www.pes-psrc.org
40. D. Spoor, J.G. Zhu, Improved single-ended traveling-wave fault-location algorithm based on experience with conventional substation transducers. IEEE Trans. Power Deliv. **21**(3), 1714–1720 (2006)
41. J. Nichols, P. Spoor, D.J. Zhu, Filtering effects of substation secondary circuits on power system traveling wave transients. Proc. Eighth Int. Conf. Electr. Mach. Syst. **3**, 2360–2365 (2005)

Chapter 9
Fault Location on Different Line Types Using Online Travelling Wave Methods

Is was shown in Chap. 4 that an impedance-based fault location method is not practically applicable for hybrid lines due to the very different fault loop impedances of these two system types. In this section, methods based on travelling waves are discussed for several line types.

9.1 Hybrid Lines

A hybrid transmission line is as mentioned a transmission line which is comprised of two or more different line types. Seen from a travelling wave fault location perspective, what makes the lines different is primarily the propagation velocity of the fault wave. On a cable system, the coaxial mode has been identified as being most suited. On an OHL, the main insulation material is air; hence, both μ_r and ϵ_r are equal to 1, and the propagation velocity of the so-called aerial mode then becomes almost equal to the speed of light (see Eq. 8.2).

A hybrid line can also be comprised of two or more cable segments with different wave velocities. The wave velocity is directly related to the relative permittivity of the main insulation and even if the same material is used in both cables, different thicknesses of the semi-conducting layers will influence the wave velocity. Assuming one constant wave velocity for such a line will result in errors if the classical two-terminal fault location method is used directly.

A hybrid fault location approach can also be used to reduce the number of necessary fault locator units at the same voltage level and thereby the cost for monitoring larger transmission systems. For instance, if three stations are connected by two lines in a radial type of configuration, fault location equipment will not be necessary in the middle station. Instead, the line can simply be seen as a hybrid line, and fault location equipment can be installed at the outer lying stations. In the coming section, a method for estimating the fault location and identifying the faulted line segment is presented.

C. F. Jensen, *Online Location of Faults on AC Cables in Underground Transmission Systems*, Springer Theses, DOI: 10.1007/978-3-319-05398-1_9,

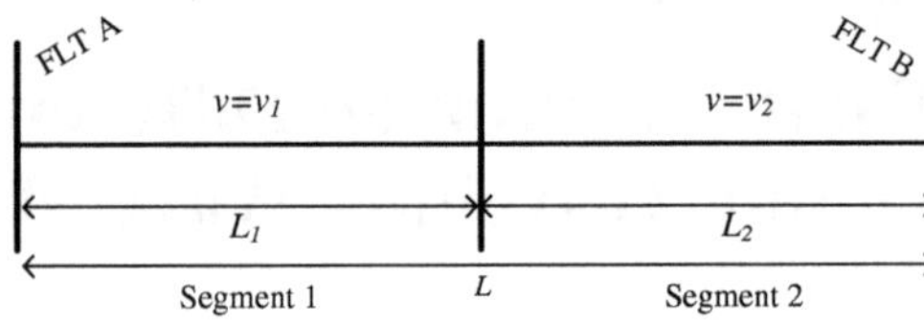

Fig. 9.1 Arbitrary hybrid line comprised of line Segment 1 with wave velocity v_1 and length A and line Segment 2 with wave velocity v_2 and length B. The total line length is L.

9.1.1 Fault Location on a Two Segment Hybrid Line

Consider an arbitrarily two-segmented hybrid line with length and wave velocities defined as shown in Fig. 9.1.

In Fig. 9.1, the wave propagation velocity of Segment 1 is v_1 and v_2 for Segment 2. The length of Segment 1 is A and B for Segment 2. The total line length is denoted L.

If a fault occurs in Segment 1 at $t = 0$, the instance the fist travelling wave arrives at Terminal A can be calculated as:

$$\tau_{A-x_1} = \frac{x_1}{v_1} \tag{9.1}$$

where x_1 is the fault location in Segment 1 seen from Terminal A. For the same fault, the first travelling wave will arrive at Terminal B at:

$$\tau_{B-x_1} = \left(\frac{L_1 - x_1}{v_1}\right) + \frac{L_2}{v_2} \tag{9.2}$$

In a fault location situation, only the time difference $\tau_{A-x_1} - \tau_{B-x_1}$ will be known. The difference between these times can be calculated as:

$$\Delta\tau_{AB,1} = \frac{L_1}{v_1} - \left(\frac{L_1 - x_1}{v_1}\right) - \frac{L_2}{v_2} = \frac{2x_1}{v_1} - \frac{L_1}{v_1} - \frac{L_2}{v_2} \tag{9.3}$$

Rearranging 9.3 in regards to x_1 will result in an expression of the fault location seen from Terminal A.

$$x_1 = \left(\Delta\tau_{A-B,I} + \frac{L_1}{v_1} + \frac{L_2}{v_2}\right) \cdot \frac{v_1}{2} \tag{9.4}$$

In Eq. 9.4, all quantities are known if the time difference $\Delta\tau_{AB,1}$ is measured using synchronised measuring equipment.

If a fault occur in Segment 2, the first travelling wave will arrive at Terminal A at:

$$\tau_{A-x_2} = \frac{x_2}{v_2} + \frac{L_1}{v_1} \tag{9.5}$$

where x_2 is the fault location on Segment 2 seen from the beginning of Segment 2. The first wave arriving at Terminal B for a fault in Segment 2 appears at:

$$\tau_{B-x_2} = \left(\frac{L_2 - x_2}{v_1}\right) \tag{9.6}$$

The difference between these times can be calculated as:

$$\Delta\tau_{AB,2} = \frac{x_2}{v_2} + \frac{L_1}{v_1} - \left(\frac{L_2 - x_2}{v_2}\right) = \frac{2x_2}{v_2} + \frac{L_1}{v_1} - \frac{L_2}{v_2} \tag{9.7}$$

Rearranging 9.7 in regards to x_2 will result in the fault location seen from the beginning of Segment 2.

$$x_2 = \left(\Delta\tau_{A-B,2} - \frac{L_1}{v_1} + \frac{L_2}{v_2}\right) \cdot \frac{v_2}{2} \tag{9.8}$$

The method can be expanded to include an unlimited number of line segments; the problem lies in determining in which section the fault occurs.

9.1.2 *Identification of the Faulted Line Segment*

When Eqs. 9.4 and 9.8 were derived, it was assumed that the faulted segment was known. This is not the case for a real fault location situation where only the time difference between the arrival instances is measured.

A method to solve this problem is proposed for hybrid DC-lines in [1]. This method can also be used on crossbonded cable systems after some modifications.

If the measured time difference $\Delta\tau_{AB,n}$ is used in both Eqs. 9.4 and 9.8, only one of the solutions will be between 0 and the length of the corresponding segment. As an example, if a fault occurs on Segment 2, the measured time difference between the waves at Terminal A and B will be given by Eq. 9.7. If Eq. 9.7 is inserted in 9.4, derived for fault in Segment 1, the result will be:

$$x_1 = \left(\frac{2x_2}{v_2} + \frac{L_1}{v_1} - \frac{L_2}{v_2} + \frac{L_1}{v_1} + \frac{L_2}{v_2}\right) \cdot \frac{v_1}{2} = x_2\frac{v_1}{v_2} + L_1 \tag{9.9}$$

Because v_1/v_2 is always positive, the calculated fault location x_1 will be estimated to be larger than A, indicating that the fault is not located in Segment 1.

Using the same approach for a fault in Segment 1, but where the fault location is estimated using Eq. 9.8, the fault location will be estimated as:

$$x_2 = (x_1 - L_1)\frac{v_2}{v_1} \tag{9.10}$$

Since v_2/v_1 is always positive and A is greater or equal x_1, the estimated fault location will be negative, indicating that the faulted section is not Segment 2. In [1], it is shown that this is the case for any fault in any segment.

9.1.3 Case Study

With the method described, it is now possible to estimate the fault location on a hybrid line including identification of the faulted segment. The coaxial mode wave velocity is used for the cable segment and the aerial mode for any OHL segment. The method is verified on a hybrid line consisting of a 3 km 1 major section crossbonded cable and a 40 km OHL which data are pretended in Appendix B. The case study 165 kV cable is used to construct the crossbonded cable (see Sect. 4.1).

Both the cable and OHL are connected to a single 400 MVA transformer with a single additional 165 kV crossbonded cable connected at the terminals. Two faults are applied to the system; A single phase to sheath fault is applied in the joint between the second and third minor section (2 km from Terminal A). Such a fault will be very difficult to lokalise using an impedance-based approach due to the very different fault loop impedances. The second fault study involves a single phase to ground fault applied 35 km from the beginning of the OHL-section. The core voltage and currents at Terminal A and B are shown in Fig. 9.2 for the two fault cases.

Inserting $\tau_{AB,1}$ and $\tau_{AB,2}$ in Eqs. 9.4 and 9.8 gives four fault locations where only two fulfill the requirement of the method. The estimated fault locations are presented in Table 9.1 with the requirement setup by the individual lines and the correct fault location underlined.

The hybrid method identifies the fault location with low error for fault in both segments; 13 m for the cable fault and 6 m for a fault in the OHL-segment.

9.1.4 Choice of Input Signal

The choice of input signal type for a fault location used on a crossbonded cable has until now been determined by the configuration of the substation. For hybrid lines, a significant discontinuity exist in the interconnection between the two line segments. A fault voltage wave travelling from the OHL section into the cable section is transmitted with a transmission coefficient of approximately 0.15. This means that the voltage wave observed at the FLT at the cable end is reduced in magnitude and, therefore, it can be difficult to extract from the power frequency signal. A current wave meeting the same discontinuity will experience a transmission coefficient of 1.85. The current wave at the FLT of the cable end is thus preferable over the voltage wave if only one additional cable is connected to the same substation; the opposite is the case for a fault in the cable section.

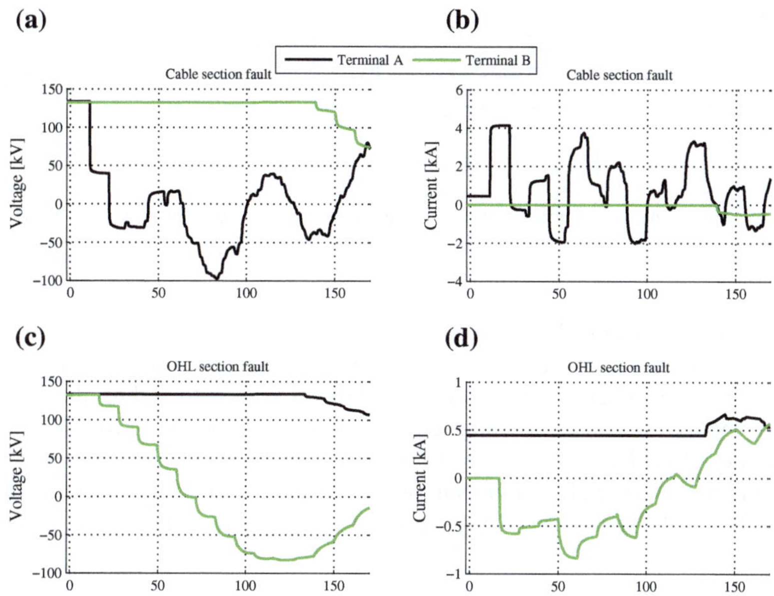

Fig. 9.2 Core voltage and current signals at the fault locator terminals for a fault in the cable section and OHL section

Table 9.1 Results for fault location on two segment hybrid line.

Eq.	Requirements	Using $\tau_{AB,1}$	Using $\tau_{AB,2}$
9.4	$0 \geq x \geq 3\,\text{km}$	2.013	24.176
9.8	$0 \geq x \geq 40\,\text{km}$	−0.163	34.994

Because the best suited signal type is dependent on in which segment the fault occurs, both voltage and currents ought to be measured in both ends of a hybrid line. More data should be gathered and a data acquisition unit with six input channels instead of three is necessary. However, one full fault locator unit is saved in the connection point between the OHL and the cable.

A three cable line segment is tested when fault location on the electrical connection to the offshore wind farm is studied in a later section.

9.2 Fault Location on Cable Systems with Solidly Grounded Sections, Transposed Cables and Cables with Open Sheath

Cable systems with one or several solidly grounded sections are being installed in the new Danish transmission system. A sketch of such a system is shown in Fig. 9.3.

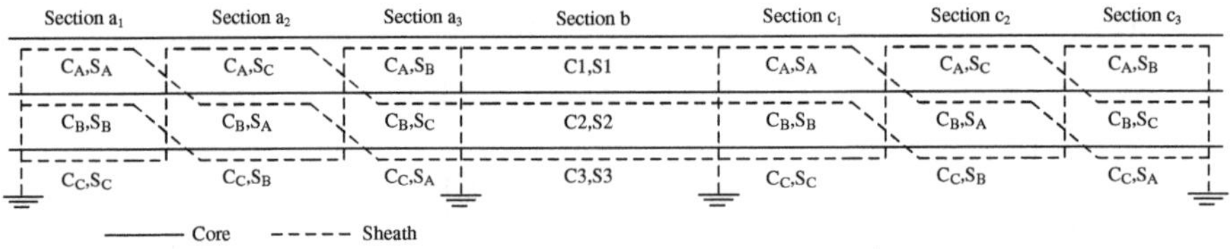

Fig. 9.3 Transmission line consisting of two major sections with a solidly bonded section in between.

A two-terminal fault location method is not affected by the solidly grounded section if coaxial waves are used as input for a fault locator. The coaxial wave will as mentioned travel between the core and the sheath closest to it; thus, the velocity is not affected by any sheath bonding scheme. There is also the case for fault location on transposed cables and cables with open sheath. However, for the latter; if the sheath is open, the sheath current cannot be measured. Instead, the sheath voltage is measurable as the sheath is no longer grounded if the core signals are not chosen.

9.3 Submarine Cables

The future Danish transmission system will include several links to offshore wind farms. Such systems are often comprised by a crossbonded land cable from the main grid to the coast wherefrom either a three-phase submarine cable or three single core submarine cables are used. For larger offshore wind parks, a common collector platform is used to connect the wind turbines.

The radially constructed link makes outage due to a fault potentially very costly, and fault location is therefore especially important on such systems.

Seen from a travelling wave perspective, the link becomes a hybrid line. It was shown in Sect. 9.1.1 that fault location on such systems is possible if a modal velocity with good characteristics for fault location (constant velocity and low attenuation) exists. Figure 9.4 shows a typical configuration of a three-phase submarine cable.

Compared to a three-phase land cable system with single core cables, the main difference is the use of an armour, the use of lead as sheath material and the fact that the cable is buried in the seabed instead of in the ground. The skin depth for the lead sheath layer is at 1 MHz 236 μm and therefore well below the thickness of the layer. As no field penetrate the sheath layer at high frequencies, the layers outside the sheath layer is of no importance. The thickness of the lead sheath layer used in the submarine cable shown in Fig. 9.4b is 2 mm. The skin depth becomes equal to 2 mm at a frequency of approximately 14 kHz. Frequency components in that range and below that penetrate the sheath layer are hence affected by other materials and the behaviour becomes hard to determine.

All submarine cables have a solid phase sheath of lead, and, hence, core/sheath coaxial modes are identified after the modal decomposition. Until better submarine

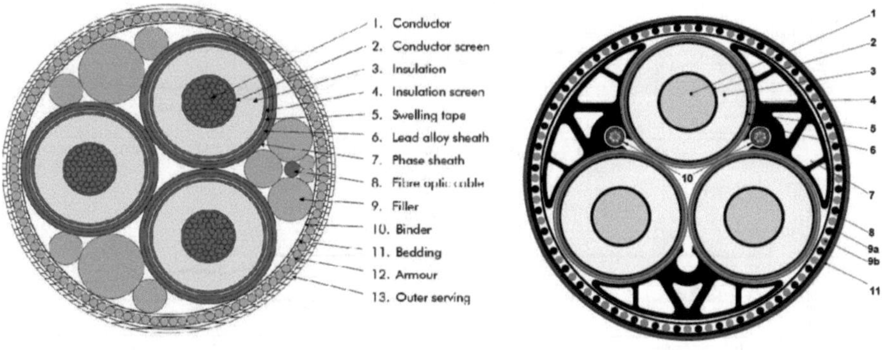

(a) (b)

Fig. 9.4 Construction of the three-phase submarine cable for **a** the Horns Reef II connection and **b** the Anholt connection. For the Anholt cable: *1* conductor, *2* inner conductive layer, *3* XLPE insulation, *4* outer conductive layer, *5* lead sheath, *6* outer PE sheath, *7* filler, *8* bedding, *9* armoring, *10* fibre optic cable and *11* outer bedding

Fig. 9.5 Single line representation of station Endrup and the Horns Reef 2 wind farm on shore connection.

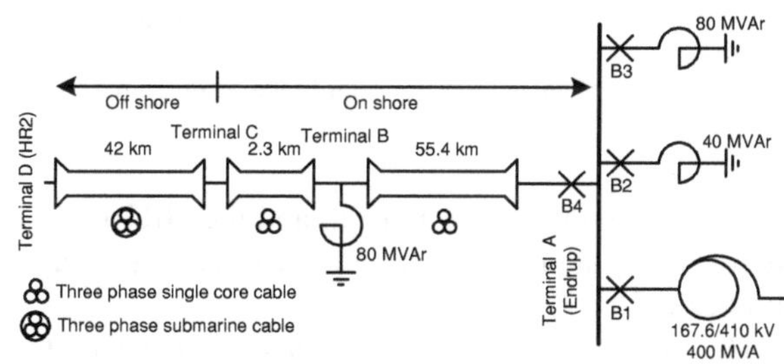

cable models are developed, a simplified version using only three coaxial type cables can be used to approximate the response at high frequencies. To examine both the simplified submarine cable model and the hybrid fault location method, a PSCSAD/EMTDC implementation of the Horns Reef 2 (HR2) electrical link is used as a case study. HR2 is a 215 MW wind farm located off the west coast of Denmark. The cable connection consists of two land cable sections with a combined length of 57.7 km and a 42 km submarine cable. To compensate for the reactive power production of the cables, a 80 MVAr shunt reactor is installed at 2.3 km from the connection point between land and submarine cable. Additionally, two reactors of 40 and 80 MVAr are installed in substation Endrup where from the power is feed to the main grid through a 400 MVA autotransformer. A single line diagram of the system is presented in Fig. 9.5.

The implementation of the two cable models is described in detail in [2].

Three different single phase to sheath faults are simulated on the cable system. Fault I is a fault in the 55.4 km land cable 31.897 km from Endrup (Terminal A), Fault 2 occurs 0.767 km in the 2.3 km land cable and Fault 3, 10 km from the intersection between the 2.3 km land cable and the submarine cable (78.7 km from Endrup). The

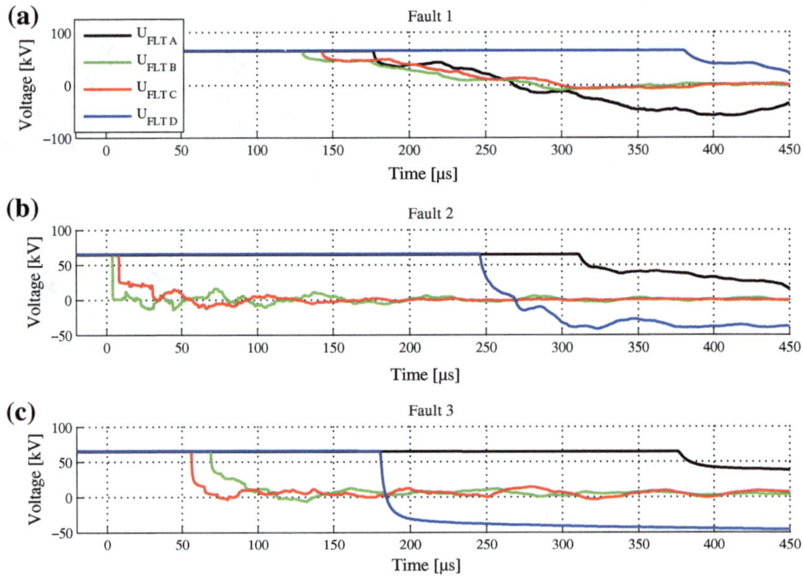

Fig. 9.6 Core voltage signals for a fault at **a** 2 km from Terminal A, **a** Fault 1, **b** Fault 2, **c** Fault 3

phase A core voltages at Endrup and the HR2 platform are shown in Fig. 9.6, and, based on the signals, the time differences $\tau_{AB,n}$ are found where n denoted fault number 1, 2 or 3.

The method described in Sect. 9.1.1 is expanded to include three line segments. For a fault in the 55.4 km land cable, the fault location can be determined as:

$$x_1 = \left(\Delta\tau_{AB,1} + \frac{L_1}{v_1} + \frac{L_2}{v_2} + \frac{L_3}{v_3}\right) \cdot \frac{v_1}{2} \tag{9.11}$$

For a fault in the 2.3 km land cable, the fault location expression is:

$$x_2 = \left(\Delta\tau_{AB,2} - \frac{L_1}{v_1} + \frac{L_2}{v_2} + \frac{L_3}{v_3}\right) \cdot \frac{v_2}{2} \tag{9.12}$$

For a fault in the submarine cable, the fault location can be determined as:

$$x_3 = \left(\Delta\tau_{AB,3} - \frac{L_1}{v_1} - \frac{L_2}{v_2} + \frac{L_3}{v_3}\right) \cdot \frac{v_3}{2} \tag{9.13}$$

Determining $\tau_{AB,1}$, $\tau_{AB,2}$ and $\tau_{AB,3}$ from a visual inspection and inserting them in Eqs. 9.11–9.13 gives six fault locations where only three meet the requirements. The estimated fault locations are presented in Table 9.2 with the requirements setup by the individual lines.

Table 9.2 Results for fault location on three segment Horns Reef 2 cable.

Eq.	Requirements	Using $\tau_{AB,1}$	Using $\tau_{AB,2}$	Using $\tau_{AB,3}$
9.11	$0 \geq x \geq 55.4\,\text{km}$	31.840	56.247	68.022
9.12	$0 \geq x \geq 2.3\,\text{km}$	−23.561	0.846	12.622
9.13	$0 \geq x \geq 42\,\text{km}$	−25.861	−0.145	10.321

Table 9.2 shows that only one of the calculated fault locations meet the requirement per fault case (underlined figure). The errors are −57, 79, and 321 m for Fault 1, 2 and 3 respectively. These errors are generally too large.

The option to install an additional fault location unit in the reactor station (Terminal B) is present on the Horns Reef 2 system. The long land part (Terminal A-Terminal B) as well as the 2.3 km land cable together with the submarine cable (Terminal B-Terminal D) can hence be monitored individually. Installing such a fault locator unit reduces the fault location estimation error for a fault in the 2.3 km section from 79 to −19 m, but more important, the error for a fault in the submarine cable is reduced from 321 to −23 m. Based on these findings, it is recommended to monitor important lines with two-terminal techniques if possible.

The work presented here, is the first to examine fault location on these special offshore connection links and is an important contribution as extended fault location on underwater cables is especially expensive.

9.4 Summary

A method capable of fault location on hybrid lines has been developed. The method uses recordings from both ends of the hybrid lines and can determine the fault location in any segment very accurately. The method is tested on a hybrid lines consisting of a 3 km crossbonded cable and a 40 km OHL as well as on the electrical connection the offshore wind farm Horns Reef 2. For long hybrid lines with submarine cables, it is recommended to install an additional fault locator unit if possible to limit the total length of the line.

9.5 Choice of Fault Location Method

Based on the knowledge obtained from the previous chapters, a suitable fault location method can now be chosen.

The velocity of a coaxial wave travelling from the fault location towards the FLT is not affected by any sheath bonding method. This is a strong advantage as all cable systems, no matter their configuration, can be monitored by the same fault locator system.

The coaxial waves are the fastest at any frequency. Due to this, the frequency dependent modal transformation is not necessary in order to extract the fault location information, and the time domain can be used directly. This is an advantage as less computational effort is required, but more importantly, the modal transformation required both core and sheath quantities as input. This means that three additional signals should be acquired for the transform to work.

The coaxial wave velocity is easy to measure in real life and it only depends on the relative permittivity of the main insulation. Because the main insulation material is homogenous along the whole cable route, the coaxial velocity is constant as well. This is a requirement for both the single and two-terminal fault location methods to work. The permittivity is not a function of temperature nor frequency and will only change slightly over time. The fact that the coaxial wave velocity is independent of any cable system parameter is another strong argument for choosing this wave type.

Due to the inductive nature of the substation's surge impedance, the voltage wave is doubled in magnitude when it reaches a substation with no additional lines connected. As more lines and especially cables are connected, the incoming core and sheath currents' waves tend to double in magnitude. The coaxial components can be equally well extracted from all signals types and the choice between them therefore becomes substation dependent. Classical capacitive voltage dividers with access to the neutral of the low voltage capacitor will provide such a measuring possibility as well. The sheath currents can be measured with low voltage high bandwidth current coils as for instance Rogowski coils.

A proposed sheath current method using the intersheath currents for a single line/single transformer station configuration is not practically applicable as it does not cover the first and last minor section of the total cable. Furthermore, the method uses the intersheath modal velocity which depends on several cable and cable system parameters. The intersheath waves are furthermore affected by the sheath bonding method and much more technical knowledge regarding wave propagation on the monitored system is needed to determine fault locations using these wave types.

The two-terminal method can be used for fault location on hybrid lines, on submarine cables, cables with solid bonded sections and cable sections with an open end. This is a strong argument for choosing the two-terminal method as the same fault locator units can monitor the entire Danish transmission system if given the right input. Based on the findings in the initial chapters of this part:

> The two-terminal method with a constant coaxial velocity as input can be used as a basic strategy to develop a fault location method for crossbonded cable systems.

However, for this statement to hold true, it must be verified that a constant coaxial wave velocity in fact can be used as input to the fault locator system and that the coaxial damping predicted in the simulations is representative for coaxial damping in real-life cables. These assumptions are verified using field measurements presented in the following chapter.

References

1. O.M.K.K. Nanayakkara, A.D. Rajapakse, R. Wachal, Location of dc line faults in conventional hvdc systems with segments of cables and overhead lines using terminal measurements. IEEE Trans. Power Delivery, **27**(1), 279–288 (2012)
2. C.F. Jensen, F. Faria da Silva, C.L. Bak, W. Wiechowski, Switching studies for the horns rev 2 wind farm main cable. Int. Conf. Power Syst. Transient (IPST) (2011)

Chapter 10
Travelling Wave-Based Field Measurements for Verification of Fault Location Methods for Crossbonded Cables

In Chap. 7, it was shown that the two-terminal method using a coaxial wave velocity with arrival instances extracted in the time domain is a good strategy for a fault locator system for crossbonded cables. However, the cable models implemented in simulations programs today does not include all layers and materials used in real-life cables and several references are studying disagreements between simulation results and field data for instance [1]. For the proposed fault location method to work, it is assumed that the coaxial wave velocity can be considered constant over the entire cable and that the coaxial attenuation is well predicted by the model; thus, the conclusions based on the simulated data so far are valid. If one of these two assumptions does not hold true, implementing a real-life two-terminal based fault locator system can be problematic. In this chapter, field measurements are carried out in order to confirm the above assumptions.

Field measurements of fault created transients are carried out on the Anholt cable; these measurements are used to verify the fault location method in general and to examine damping of the fault waves. Because the coaxial wave velocity is an input parameter for the proposed online fault location method, different ways of measuring this velocity on real-life crossbonded cables are proposed. This is done through a series of impulse measurements where these are used to study the effect of coaxial damping on a real cable system as well.

First, the measuring strategy together with the measuring equipment used is described, then the modal decomposition of the Anholt cable is determined. The modal decomposition is interesting as an ECC which is installed with an unknown position. The effect on the transient response of this additional conductor is studied as well. A PSCAD/EMTDC simulation model is constructed for both the 'travelling wave-based fault location measurement' and impulse studies. The simulations are used to help select the correct measuring equipment and to study general transients behaviour of the Anholt system. Impulse measurements are carried out on the 6.9 and a 31.4 km Anholt cable sections described earlier in this report. In the final section, four case studies are presented showing the possible use of the two-terminal method for fault location on crossbonded cables.

C. F. Jensen, *Online Location of Faults on AC Cables in Underground Transmission Systems*, Springer Theses, DOI: 10.1007/978-3-319-05398-1_10,

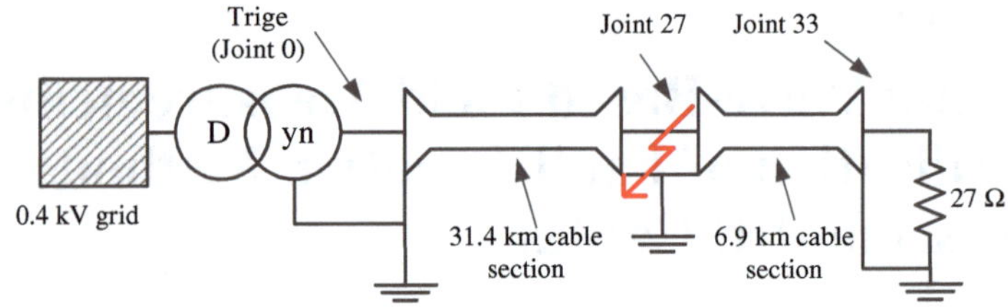

Fig. 10.1 Sketch of the 245 kV crossbonded cable system under study

10.1 Measuring Strategy

The crossbonded Anholt land cable section described in Sect. 5.1.1 is also used for travelling wave-based fault location measurement. Using the same strategy as for the impedance based measurements, fault location on cable systems with different lengths is possible (see Fig. 5.7). This can help confirm or reject the claim that fault location using a constant coaxial wave velocity is a good solution for different cable lengths. Four case studies are set up and will be described in Sect. 10.5.

In order to emulate a fault representing an insulation breakdown as good as possible, a switch with reduced pre-strike capability is used. A Siemens 3AF 1532-4, 12 kV 1230 A vacuum circuit breaker was available and connected between the core and the three bonded and grounded sheaths at Joint 27. The fault inception angle could not be controlled using the vacuum breaker, wherefore, instead, the breaker was closed at several arbitrarily instances, the most interesting recorded, and, later the fault inception angle in the simulation was adjusted accordingly.

All measurements were performed at reduced voltage (0.4 kV) under the assumption that the voltage/current relationship of the cable system is linear, so the travelling waves will be representative for travelling waves created during a real-life fault situation. The core voltages, core currents and sheath currents are all of interest for fault location purposes and are thus measured. A 10 kVA 0.4/0.4 kV Dyn transformer was used to feed the system. The use of a transformer gives optimal conditions for detecting the voltage waves as the incident fault wave is fully reflected and therefore doubled. The core and sheath currents at the transformer location are not attractive due to the behaviour that occurs when a wave meets the inductance of the power transformer.

It was not possible to connect another high voltage cable at the far-end compared to the transformer location. Instead, a load resistor of 27 Ω is used. This resistor represents a reflection less cable line with approximately the same surge impedance as the Anholt cable. This gives equal condition for capturing fault waves in both the core voltage, core current and sheath current as only a fraction of the incident waves is reflected. The use of a load resistor will also ensure a steady state current flowing prior to fault. Figure 10.1 shows the principle for the travelling wave-based measurements for energisation at Joint 0, the fault location at Joint 27 and the load resistor at Joint 33. Depending on the case study performed the location of the components can be interchanged.

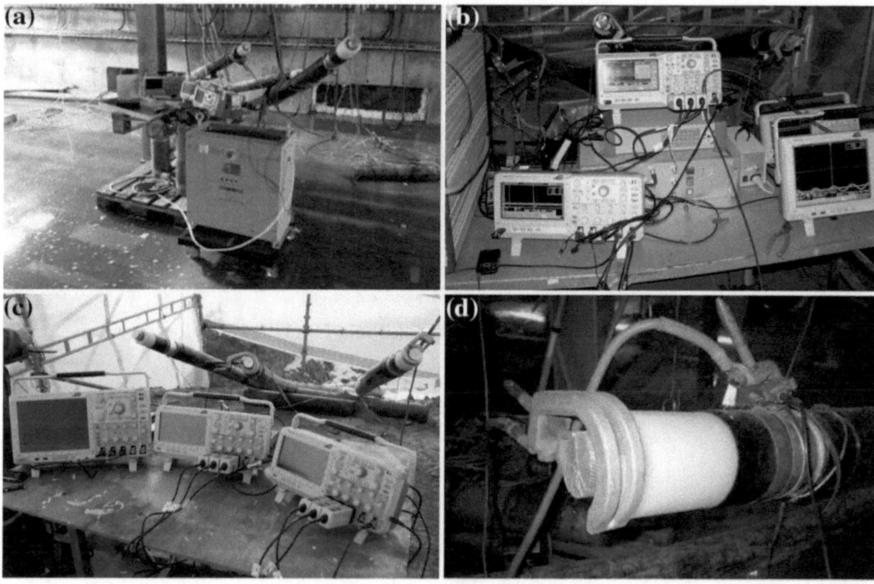

Fig. 10.2 Pictures from the measurements. **a** The setup in Substation Trige, **b** at Joint 27, **c** at Joint 33 and **d** connection of a cable core and sheath for measuring purposes

Table 10.1 Voltage and current probes used for field measurements

Fabricate and ID	Type	Bandwidth (MHz)	Range
Tektronix P5205A	Voltage differential probe	100	± 1300 V
Tektronix TCP0030	Current probe	120	30 A RMS/50 A peak
Tektronix TCP0150	Current probe	20	150 A/500 A peak

Pictures from the different measuring locations are shown in Fig. 10.2.

Voltage and Current Probes

High bandwidth measuring equipment was used to measure the voltage and current signals at the cable system ends after the breaker was closed. The sheath voltages were assumed to be zero relative to neutral ground at all locations as measurements of local voltage rises were not possible.

The voltage and current probes were selected prior to the field measurements based on simulations conducted in the transients simulation program PSCAD/EMTDC. The results of these simulations and the setup of the simulation model are presented together with the results from the field later in this chapter. Based on the simulation results, the probes shown in Table 10.1 were selected.

As seen from Table 10.1, the lowest bandwidth of the used current probes was 20 and 100 MHz for the voltage probes. The voltage and current signals are measured directly on the conductors, and the effect of power system instrument transformers are thereby neglected.

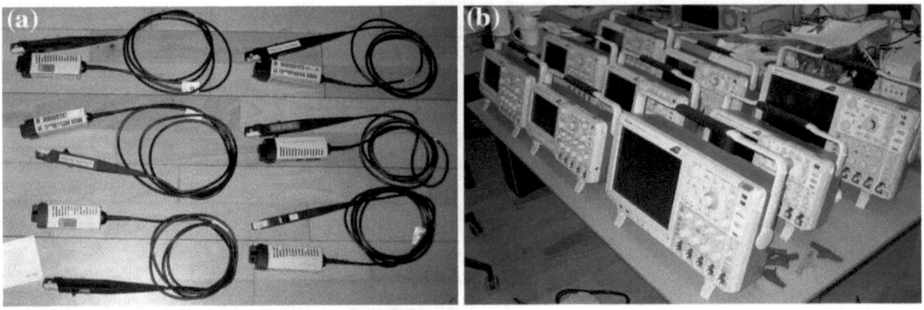

Fig. 10.3 **a** Tektronix TCP0030 current probes. **b** Tektronix DPO2014 oscilloscopes

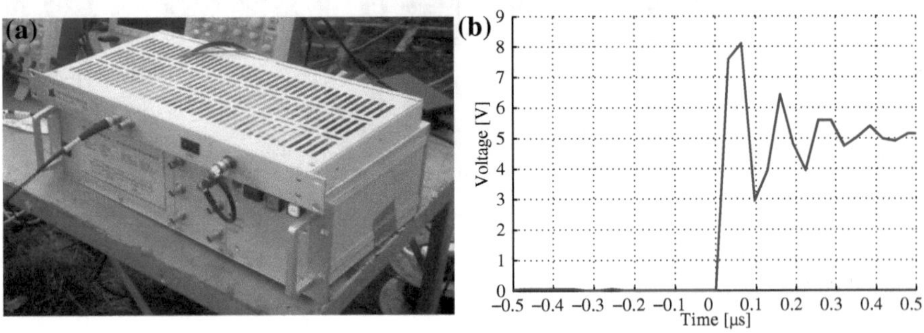

Fig. 10.4 **a** Rubidium-based synchronisation unit at measuring location. **b** Output signal at the instance a pulse is created

Data recording units The signals were recorded with Tektronix DPO2014 oscilloscopes set at a sampling frequency of 31.25 MHz and the vertical resolution of the oscilloscopes was 8 bits or 256 points. In Fig. 10.3a, some of the Tektronix TCP0030 current probes used are shown, and in Fig. 10.3b, several Tektronix DPO2014 oscilloscopes.

Time synchronisation In order to synchronise the measured data, a highly accurate common time reference was needed at both cable ends. To provide this, synchronisation units based on the 'Rubidium Frequency Standard' were used [2]. These units generate a square wave signal very well defined in time every 40 ms. This pulse is recoded using a channel on each oscilloscope providing a time reference common for all oscilloscopes.

The synchronisation units are physically brought together before the measurements are started and synchronised. Subsequently, they are taken to the measuring location powered by battery. In Fig. 10.4a, a unit setup at Joint 33 is seen, and in Fig. 10.4b, the measured output is shown.

Figure 10.4b shows that the instance of pulse is easily identified. The error is maximum one sample delay ($1/f_s = 32$ ns).

Once the measurements had been carried out, the synchronisation units were brought back together to examine if they had drifted in time. The synchronisation units drifted a maximum of 20 ns. This is below one sample delay, hence, no effort was made to correct for the error.

10.1.1 Equipment Accuracy

The main purpose of the field measurement carried out in this thesis is to determine the arrival instance of fault created travelling waves. Therefore, the horizontal time resolution of the data recording units is of importance. The sample frequency of the Tektronix DPO2014 oscilloscopes is set to 31.25 MHz. Disregarding the sample rate accuracy, any incoming wave can be determined with an accuracy of 32 ns; this is equivalent to a coaxial wave travelling 5.7 m. Using the method presented in Sect. 8.9 to determined the estimation error for the two-terminal method as function of sampling frequency, a worst case error of ± 2.9 m is determined.

The sample rate accuracy is ± 25 ppm for intervals larger than 1 ms. This deviation can be disregarded totally due to the small error introduced.

10.2 Modal Decomposition of the Anholt Land Cable Section

The modal decomposition of the Anholt cable is different compared to the cable system laid in flat formation discussed in Sect. 6.1. Due to the ECC, an additional mode is introduced and the characteristics of this mode must be studied.

The ECC is most likely placed very close to phase A wherefore a high intersheath modal velocity is expected due to the small loop between the ECC and the sheath of cable A. It is assumed for the proposed fault location method that the coaxial waves are the fastest at any frequency. If the new intersheath mode for any placement of the ECC can become faster than the coaxial waves, the proposed fault location methods will not be applicable. Whether or not this is the case is the main focus of this section.

After the modal decomposition, a three single core cable system is decomposed into three coaxial, two intersheath and a ground mode. Including an ECC adds an additional mode to the system as the 6×6 series impedance and shunt admittance matrices becomes 7×7 matrices. The ECC is, as mentioned in Sect. 5.5, electrically disconnected at Joint 6, 12, 18, 24 and 33. The ECC will, however, still affect the transient response of the Anholt system due to magnetic coupling and must therefore be included in the modal calculations and when constructing the simulation model.

The only knowledge regarding the position of the ECC is that it is laid with phase A the entire cable route. In comparison to the cable sheaths, and especially the sheath of phase A, the position of the ECC can change relatively much.

The frequency dependent 7×7 transformation matrix $\mathbf{Q}$, calculated for the Anholt cable system reveals how the modes are affected by the ECC. Assuming the correct ECC position, evaluating $\mathbf{Q}$ at 1 MHz and taking the real part (neglectable imaginary parts) yields the result seen in Eq. 10.1.

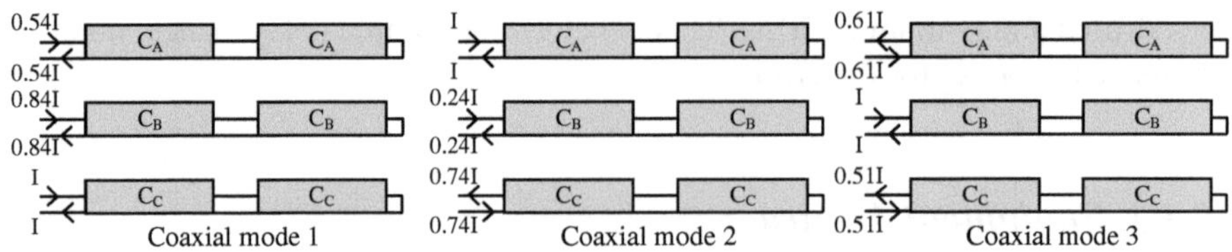

Fig. 10.5 Visual presentation of the three coaxial mode currents used for the Anholt cable

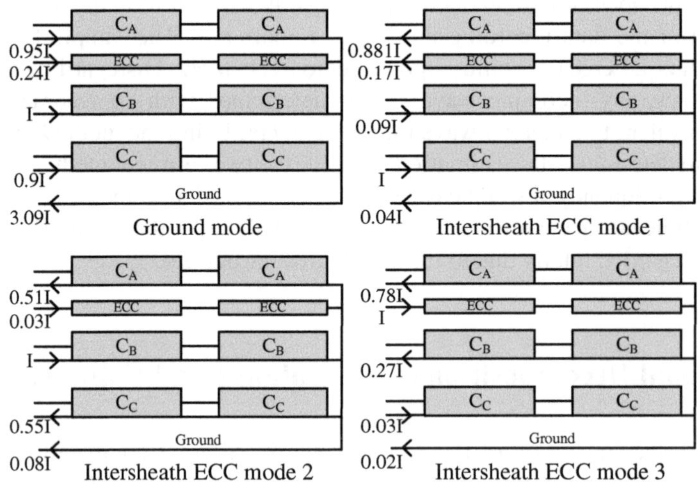

Fig. 10.6 Visual presentation of the ground and three intersheath ECC modes for the Anholt cable

$$\mathbf{Q} = \begin{bmatrix} 0 & 0 & 0 & 0 & 0.54 & 1.00 & -0.61 \\ 0.95 & -0.88 & -0.50 & -0.78 & -0.5400 & -1.00 & 0.61 \\ 0 & 0 & 0 & 0 & 0.84 & 0.24 & 1.00 \\ 1.00 & 0.09 & 1.00 & -0.27 & -0.8400 & -0.24 & -1.00 \\ 0 & 0 & 0 & 0 & 1.00 & -0.74 & -0.51 \\ 0.90 & 1.00 & -0.55 & 0.03 & -1.0000 & 0.74 & 0.51 \\ 0.24 & -0.17 & -0.03 & 1.00 & 0 & 0 & 0 \end{bmatrix} \tag{10.1}$$

A visual interpretation of the three coaxial modes defined in column 5–7 of **Q** are presented in Fig. 10.5.

Columns 5–7 in **Q** shows three coaxial modes with all current propagating between the cores and sheaths. The coaxial modes are, as expected, not affected by the ECC at high frequencies as the cable sheaths provide perfect shielding. The ground and three intersheath modes defined by columns 1–4 are, however, strongly affected as indicated in Fig. 10.6.

Modal current will flow in the ECC for all four non-coaxial modes and it is noticed that the ground return is excited for all modes. This is the case for all unsymmetrical laid three single-core cables, but the effect is more pronounced when the ECC is included. Mode 1 (Column 1 in **Q**) is a classical ground mode with all current returning in the ground where mode 2–4 (Columns 2–4 in **Q**) will be considered 'Intersheath ECC modes' due to the limited excitation of the earth. Because of the high percentage of the modal current flowing in the ECC for intersheath ECC modes 1, but especially

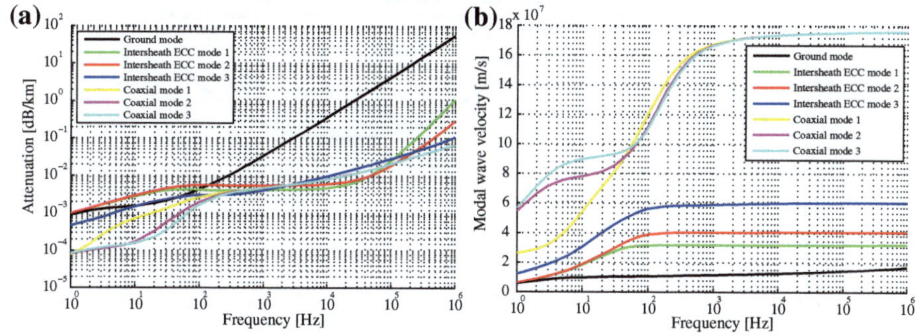

Fig. 10.7 **a** Modal attenuation. **b** Velocity as a function of frequency three cable system laid in flat formation with an ECC

Table 10.2 Modal velocities evaluated at 1 MHz

Mode	Velocity [m/s]
Coaxial mode 1–3	$1.7615 \cdot 10^8$
Ground mode	$0.1688 \cdot 10^8$
Intersheath ECC mode 1	$0.3129 \cdot 10^8$
Intersheath ECC mode 2	$0.4025 \cdot 10^8$
Intersheath ECC mode 3	$0.6064 \cdot 10^8$

mode 3, these modal velocities will be dependent on the position and parameters and placement of this conductor.

The modal velocities as function of frequency are calculated from 1 Hz to 1 MHz for the first minor section, and the result is presented in Fig. 10.7.

The coaxial modes show the same behaviour as for a classical three single core cable laid in flat formation at high frequencies. The intersheath ECC modal velocities are very different from each other. The current flows for Intersheath ECC mode 1 primary between the sheaths of cable A and C. For Intersheath ECC mode 2, the primary current flows into the sheath of the middle cable and returns in the sheaths of the two others. Intersheath ECC mode 3 is the fastest as most current flows between the sheath of cable A and the ECC placed 0.1333 m from it. The current loop is thus small resulting in a smaller inductance and hence a higher velocity. The velocity of intersheath ECC mode 1 and 2 are frequency independent above 100 Hz where intersheath ECC mode 3 becomes frequency independent after 1 kHz. The result is five unique modal velocities in the system at high frequencies. The modal velocities evaluated on the Anholt cable at 1 MHz are presented in Table 10.2.

Because the ECC is placed with phase A on the entire cable route, the modal decomposition of each of the minor section will result in the same velocities, and further calculations are unnecessary. For a healthy system, the ECC is mirrored about phase B each time the ECC is shifted, thus the same physical distances exist between the conductors and the same modal velocities are found after the decomposition—only their definitions are affected.

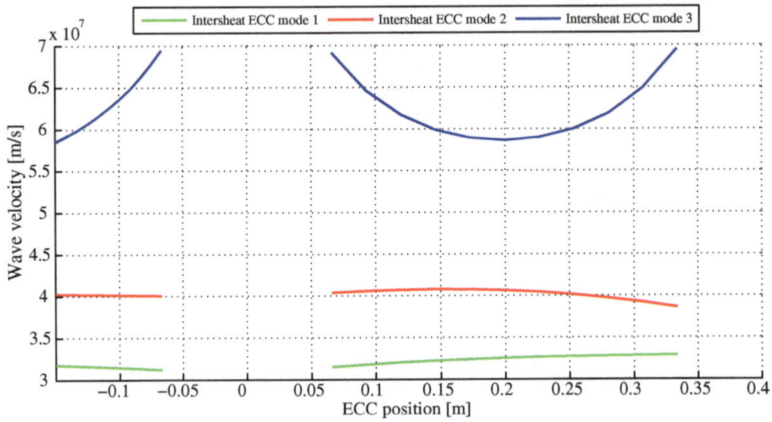

Fig. 10.8 Sensitivity of the intersheath ECC mode velocities on the position of the ECC

As shown in Table 10.2, the Intersheath ECC mode with characteristics determined mostly by the ECC/Sheath A circuit is the fastest of the intersheath ECC waves. As the ECC is moved closer to phase A, it is expected that the velocity of this mode will increase. If the velocity becomes comparable to the coaxial velocities, problems with the initial assumption of the proposed fault location method can arise. In the following section, this is examined.

10.2.1 The Influence of the Position of the ECC on the Modal Velocity

The position of the ECC in the Anholt cable system is not precisely known. Due to installation issues, it was laid with Cable A on the entire route. However, no attempt was made to ensure a fixed position of the ECC which means that it can be placed more or less close to cable B or on the far side of Cable A. The Intersheath ECC modes are on the other hand strongly affected as the size of the loop formed by the cables' sheaths, and the ECC will change as the position of the ECC changes. Assuming a fixed depth of the ECC, the Intersheath ECC modal velocities are calculated as the cable is moved from −15 cm to cable A and then again from the other side of cable A until it touches cable B. The result is shown in Fig. 10.8.

As expected, the position of the ECC influences the Intersheath ECC 3 mode velocity strongly. When the ECC forms a small loop with either the sheath of cable A or B, the velocity is at its highest. At a position of 0.2 m, the ECC is precisely between cable A and B, and the lowest velocity is observed. The other two intersheath ECC modes are dependent as well, but to a lesser degree.

The velocity of the Intersheath ECC modes will in fact influence the results of the field measurements, but not the ability to use coaxial waves velocity as an input for a

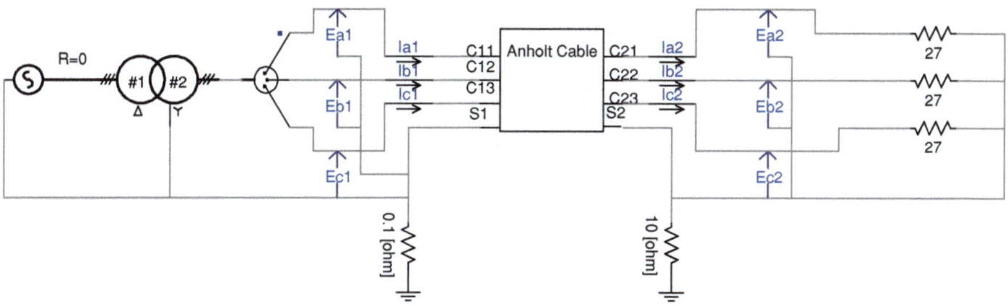

Fig. 10.9 PSCAD/EMTDC implementation of the land cable section from Joint 0 to 33 of the Anholt cable system

two-terminal fault locator system. Because the position is unknown and most likely will change along the cable route, uncertainties are introduced in the simulations and theoretical calculations. However, the coaxial waves will always arrive first and as these are not influenced by the position of the ECC, no errors are introduced by the ECC until the intersheath ECC modal waves begin to arrive. As the proposed fault locator system is preferably based on coaxial modal waves, the ECC issue is not of importance to the accentual fault location results. General simulation studies conducted on the Anholt system should therefore be carried out with care.

10.3 Simulation Model

A model of the Anholt cable system from Joint 0 to 33 is implemented in the transient simulation program PSCAD/EMTDC. The implementation is seen in Fig. 10.9.

The cable sections are modelled using the 'Frequency Dependent (phase) Model' which is based on the Universal Line model by Gustavsen [3–6]. The parameters used in the model are calculated using the CABLE CONSTANTS subroutine available in PSCAD/EMTDC [7]. The CABLE CONSTANTS subroutine requires parameters based on a geometrical description of the cables and their position. Besides the physical size of the layers and their positions, the subroutine requires a number of material constants that describes the electrical characteristics of the materials used (resistivity, permittivity, permeability etc.). The core and sheath resistivities and the main insulation permittivity are corrected using the methods presented in [8, 9]. The ECC is modelled as a solid core cable with a polyethylene (PE) insulation with a permittivity of 2.3. The conductor is placed next to cable C on the entire route and electrically disconnected at Joint 6, 12, 18, 24 and 33. In Fig. 10.10, the PSCAD/EMTDC implementation of the cables' model is shown.

The material constants after correction are presented in Table 10.3.

In Case Study II (described later in this chapter), an ideal resistor is used to model both the load resistor and the fault resistor. The two-winding transformer model available in the PSCAD/EMTDC library is used to model the 0.4/0.4 kV transformer

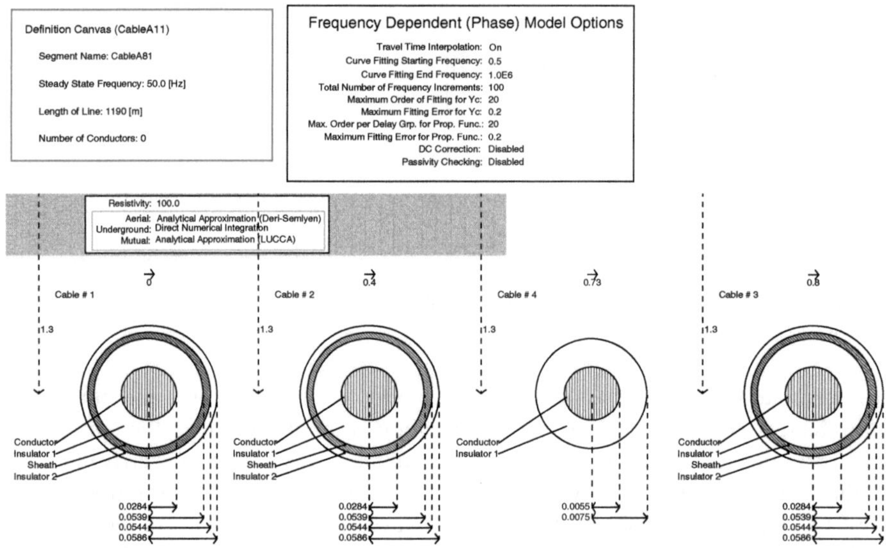

Fig. 10.10 PSCAD/EMTDC implementation of the land cable section from Joint 0 to 33 of the Anholt cable system

Table 10.3 PSCAD cable model parameters for the Anholt cable

Parameter	Value
ϵ_{i1}	2.889
ϵ_{ECC}	2.3
ρ_c	$3.5474 \cdot 10^{-8}\ \Omega \cdot \text{m}$
ρ_s	$2.6756 \cdot 10^{-8}\ \Omega \cdot \text{m}$
ρ_{ECC}	$1.724 \cdot 10^{-8}\ \Omega \cdot \text{m}$
ρ_g	$100\ \Omega \cdot \text{m}$

neglecting any stray capacitance. For high frequency studies, the grid behind the transformer is unimportant and is therefore not considered in the model.

The grounding resistances at the crossbonding locations are modelled as 10 Ω resistances whereas the grounding resistance at substation Trige is set to 0.1 Ω.

The time step is set to 0.031 μs equal to the sampling delay used for performing the measurement. All simulations are set to run until steady state is reached before a fault is applied. The fault inception angle is fitted to the measured angle based on a visual inspection. All results are presented in the following sections including the measuring results.

For determining the coaxial wave velocity, impulse measurements are used. The impulse generator is modelled using two exponential curves and a resistance equal to that of the surge generator (40 Ω).

10.4 Coaxial Wave Velocity Determination

The coaxial wave velocity is a necessary input parameter for all travelling wave-based fault location methods. Therefore, methods for determining this velocity must be considered. Furthermore, whether it is correctly to assume a constant velocity for cable systems of different length as well as the coaxial damping in the system must be evaluated.

As discussed, three unique coaxial modal signals travel with the same velocity at frequencies above approximately 1 kHz and the velocity becomes frequency independent above 10 kHz on the Anholt cable (see Fig. 10.7b). As these types of waves are the fastest at any frequency, the first signal change noticed at any location will be due to a coaxial wave. As described in Sect. 6.2, a coaxial wave is not affected by the crossbonding of the sheath. Because of the definition of the coaxial wave, a suitable method to emulate this wave type is applying an impulse between the core and the sheath. It is assumed that the coaxial velocity can be considered constant. Impulses are applied at Joint 27 and the response measured at both Joint 0 and 33. The determined velocity for waves which have travelled 6.9 and 31.4 km can be compared and the claim tested. Three methods are proposed for evaluating the coaxial modal velocity. These are:

- **The Classical Time-domain Reflectometer Method (Method 1)** can be difficult to use on long crossbonded cables. This is due to the reflections created at each crossbonding. Because of these additional reflections compared to an OHL, it can be difficult to identify which reflection has travelled from the receiving end and thereby which reflection should be used to calculate the wave velocity. On shorter cables, this reflection stands out in such a way that it can be identified if the termination of the receiving end is chosen wisely.
- **The Crossbonding Created Wave Method (Method 2)** can be used to determine the wave velocity on a cable under the assumption that the velocity is constant over the entire cable route. When the wave sent into the cable meets a crossbonding, it is divided into a reflected and refracted wave. If the reflected wave is measured at the sending end and the length is known, the wave velocity for that section can be determined. The method can be used arbitrarily to include any number of minor sections as long as the waves can be determined at the sending end. If the first reflection is used, only the velocity on the first minor section is determined. If the second reflection is used, the calculated velocity will be a mean value of the velocity in the first two minor sections and so forth. The method is robust as it requires no synchronisation equipment.
- **The First Wave at the Receiving End Method (Method 3)** will provide the most accurate estimate of the propagation delay. The first wave arriving at the receiving end is coaxial of nature and the wave velocity will simply be l_c/τ_d where l_c is the total cable length and τ_d is the wave's travelling time from sending to receiving end. The method is very precise, but difficult to use in practice because of the very accurate time synchronisation required.

Table 10.4 High voltage probes used for impulse measurements

Fabricate and ID	Type	Bandwidth	Range (V)
Sapphire SI-9010	Voltage differential probe	DC-70 MHz	± 7000
Pintex DP-50	Voltage differential probe	DC-50 HHz	6500

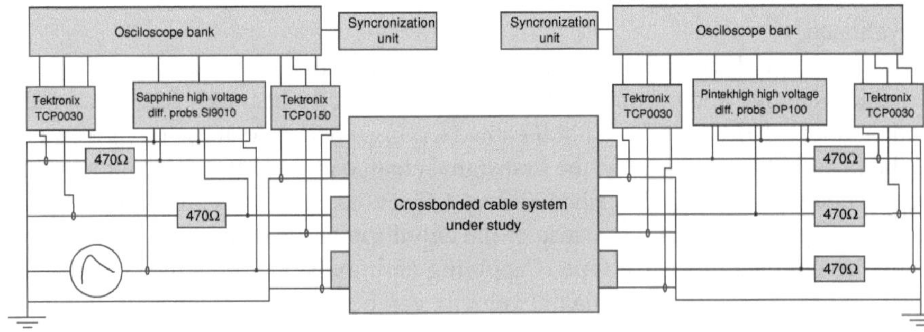

Fig. 10.11 Measuring system used for the lightning impulse field measurements performed in the case of a phase energisation

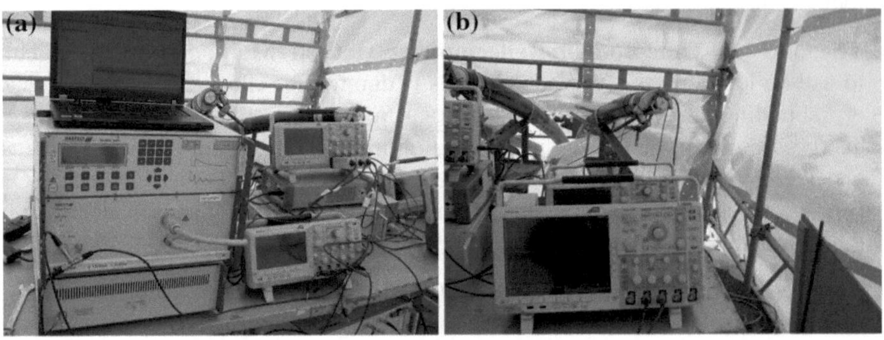

Fig. 10.12 Measuring setup at Joint 27 for impulse measurements on the Anholt cable section from Joint 27 to 33

The impulse applied to the cable system was generated using a HAEFELY PSurge 8000 surge generator. The three core voltages and currents as well as the three sheath currents were measured at both cable ends. The sheath voltages were assumed to be zero relative to neutral ground. The non-energised cores as well as all cores in the receiving end were grounded through 470 Ω resistances in order to give measurable currents. The same data acquisition, the synchronisation unit and the same high bandwidth current probes as described in Sect. 10.1 were used. For measuring the high voltages, the voltage probes shown in Table 10.4 were used.

The sampling frequency of the Tektronix DPO2014 oscilloscopes was set to 31.25 MHz. Figure 10.11 shows the measuring setup for a single phase energisation.

An example of the measuring setup at is shown in Fig. 10.12.

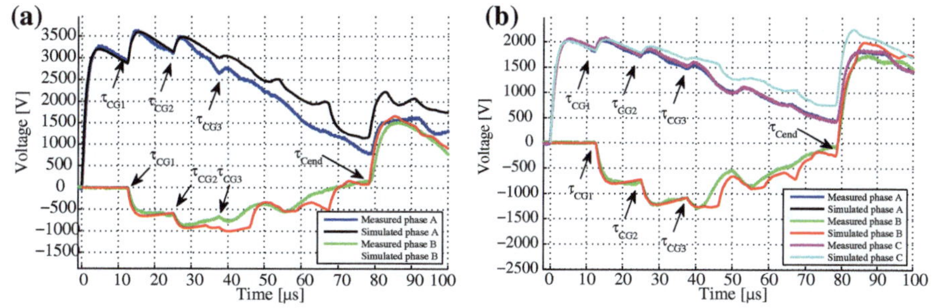

Fig. 10.13 Measured and simulated core voltages at Joint 27 for the energisation of **a** phase A, and **b** Phase A and C at the sending end of the 6.90 km Anholt cable section

Short cable section Field measurements of energisation of all possible combinations of phase conductors are carried out on the 6.905 km cable section from Joint 27 to 33 (single phase, two phase and three-phase). The same simulation model as described in Sect. 10.3 is used, but with a model of the surge generator connected at Joint 27. The magnitude of the simulated impulse is adjusted to fit the measured value.

The *n*th coaxial wave group arriving at the joints is denoted τ_{CG_n} both at the sending and receiving end. In the following, some arbitrarily chosen measurements are shown along with the simulated signals.

In Fig. 10.13a, the sending end phase A and B voltage for a phase A energisation is shown. Figure 10.13b shows the sending end voltage of all three phases for a phase A and C energisation (two-phase).

It is seen that all instances marked in the figures are identifiable. As more reflections arrive, the exact instance becomes harder to determine due to damping. The arrival of the first wave from the first crossbonding (τ_{CG_1}) can be determined very accurately, but it represents only the travelling time in the first minor section. The instances τ_{CG_2} and τ_{CG_3} contain information about the travelling time of the first two and three minor sections, but are harder to determine accurately. The first reflection recorded at the sending end travelling from the receiving end can be determined with quite high accuracy ($\tau_{CG_{end}}$). This reflection is the most suited when using sending end reflections if it can be determined correctly as it represents an average velocity for all sections from Joint 27 to 33.

The results presented in Fig. 10.13a and b shows that the simulation model predicts the level of attenuation well as long as only coaxial waves contribute to the voltage profiles. As more waves and especially intersheath ECC waves begins to arrive, the match between simulation and measurement lowers. As only coaxial waves are necessary for the proposed fault location method to work, the deviation are less important in this case.

The phase voltages and currents at Joint 33 in case of a phase A energisation at Joint 27 are shown in Fig. 10.14a and b, respectively.

Again, the simulated and measured signals are very comparable as long as only coaxial waves contribute to the signals indicating that the cable model used performs

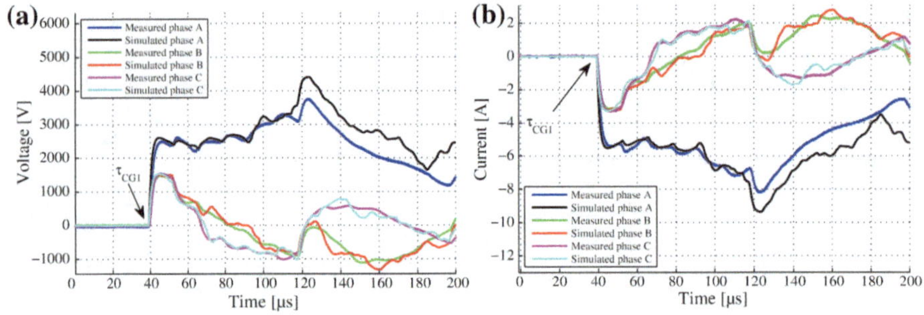

Fig. 10.14 **a** Voltages and **b** currents at the receiving end (Joint 33) of the 6.90 km Anholt cable section for a Phase A energisation at Joint 27

Table 10.5 Mean value of the coaxial wave velocity determined on a 6.9 km crossbonded cable

Method	Reflection	Time [μs]	Length [m]	Velocity [m/s]
1	$\tau_{CG_{end},S}$	79.69	13810	$1.7489 \cdot 10^8$
2a	$\tau_{CG1,S}$	12.68	2206	$1.7428 \cdot 10^8$
2b	$\tau_{CG2,S}$	25.35	4412	$1.7404 \cdot 10^8$
3	$\tau_{CG1,R}$	39.61	6905	$1.7470 \cdot 10^8$

well. The unknown location of the ECC is expected to contribute to the error as the Intersheath ECC waves start to arrive. In general, the simulation shows a lack of damping which is a common observation and discussed in detail in several publications [1, 8, 10, 11]. As the coaxial waves are well predicted, it can be concluded that the representation of the cable models in PSCAD/EMTDC can be trusted when the signals are only used for two-terminal fault location determination. The arrival instance of the first wave at Joint 33 is easy to pick up and the wave velocity over the entire cable can be determined with high accuracy.

From analysing all the measured voltage and currents signals for all measurements carried out on the Joint 27 to Joint 33 cable section, it is found that the coaxial travelling time is almost determined identical when determining it from the voltage and current signals under the different phase energisation schemes (τ_{CG_1} in Fig. 10.14a is equal τ_{CG_1} in Fig. 10.14b and so forth). Table 10.5 presents a mean value of all measured coaxial wave velocities determined using the TDR Method (Method 1), The Crossbonding Created Wave Method including the use of both τ_{CG_1} and τ_{CG_2} (Method 2a and b), and The First Wave at The Receiving End Method (Method 3).

It is noticed that the calculated velocity based on the simple relationship seen in Eq. 8.2) is very accurate; 0.13 % compared to the velocity found using Method 3.

Long cable section The same measurements are all carried out on the 31.4 km long crossbonded cable system from Joint 0 to 27 with the system energised from Joint 27. As an example, the phase A voltage and current at Joint 27 in case of a phase A energisation are shown in Fig. 10.15a and b and the phase A voltage and current for the same configuration at Joint 33 are shown in Fig. 10.16a and b.

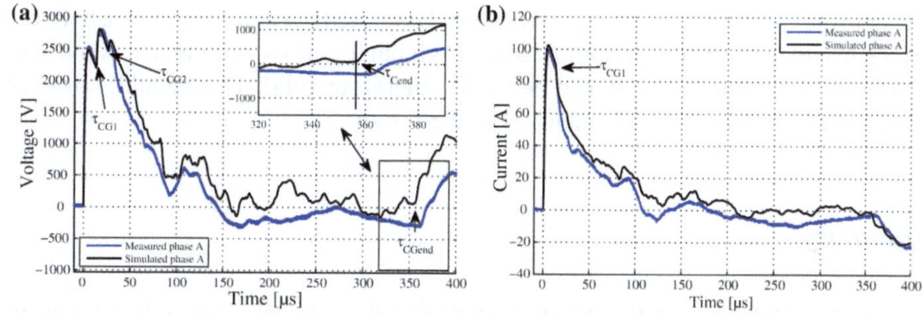

Fig. 10.15 **a** Voltages and **b** current at the sending end of the 31.4 km Anholt cable section for a Phase A energisation at Joint 27

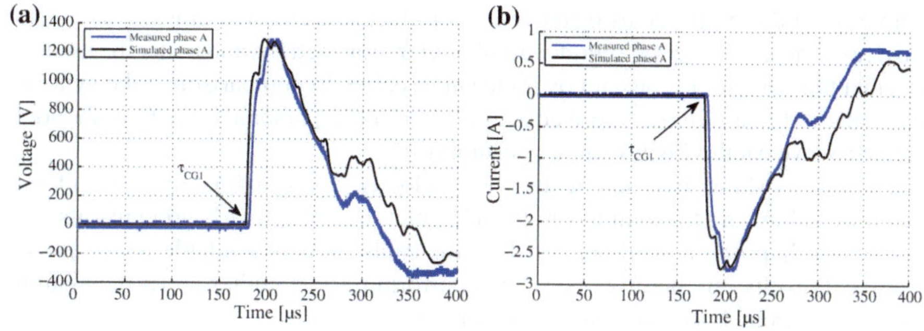

Fig. 10.16 **a** Voltages and **b** current at the receiving end of the 31.4 km Anholt cable section for a Phase A energisation at Joint 27

Table 10.6 Coaxial wave velocity determined on a 31.4 km crossbonded cable

Method	Reflection	Time [μs]	Length [m]	Velocity [m/s]
1	$\tau_{CG_{end},S}$	362.6	62800	$1.7327 \cdot 10^8$
2a	$\tau_{CG1,S}$	12.76	2222	$1.7311 \cdot 10^8$
2b	$\tau_{CG2,S}$	25.48	4444	$1.7343 \cdot 10^8$
3	$\tau_{CG1,R}$	181.16	31400	$1.7333 \cdot 10^8$

Figures 10.15 and 10.16 show that the measured signal in general has a higher degree of damping compared to the simulated signals. All arrival instances are, however, determined with success. The resulting coaxial wave velocities are presented in Table 10.6.

The results show that single-ended measurement methods can be used to determine the coaxial wave velocity even on long crossbonded cables and thereby reduce the cost and complexity of the measurements compared to using synchronised measurements. The velocities determined on the 6.9 and 31.4 km sections are almost equal with the highest velocity on the short section as expected (0.39 % error between the velocities determined on the two sections using Method 3). Assuming a constant velocity will introduce the largest error for a fault close to one FLT. As the fault location moves closer to the cable middle, the error introduced is evened out as the difference between arrival instances at the two terminals is used as input for the two-terminal method. The claim that a constant velocity can be used is therefore valid. This should, however, be tested as more field data is gathered in the future.

Summary

The results presented in this section show that the coaxial wave velocity can be determined quite easily using several different methods.

A synchronised measuring setup (The First Wave at The Receiving End Method) will give a representation of the coaxial wave velocity over the entire cable. A more simple method is proposed using the reflections from the crossbondings. This method provides a mean value over a certain number of minor sections and under the assumption that the velocity is constant over the entire cable. In comparison to the synchronised method that requires more equipment and more personnel, the Crossbonding Created Wave Method is less time consuming.

The results show that the coaxial wave velocity is approximately constant over the entire cable, and the simple single ended method will provide good results.

Coaxial damping is well predicted by the cable models available today and the simulated results can therefore be used to perform further studies under the assumption that the model used is constructed correct.

10.5 Case Study Results

In this section, field measurements which purpose are to verify the proposed fault location method for crossbonded cables are presented. A total of fifteen measuring sets were recorded on the Anholt land cable part between either Joint 0 and 33, between Joint 0 and 27 or between Joint 27 and 33 (a measuring set is the recording of all core voltages and currents as well as the sheath currents). Only single phase to sheath to ground faults are emulated. The closure of the breaker's poles occurs with a time delay determined by the mechanical properties of the breaker. This time delay is many times larger compacted to the time it take an arc to establish a connection between the faulted conductors. In the following sections, four of the most interesting of these measuring sets are presented. The definition of the case studies is shown in Table 10.7.

Case Study I represents an insulation breakdown near negative peak voltage occurring at Joint 27 with measuring locations at Joint 0 and 33 (38.3 km cable). This is a common case, where the insulations break down under high stress. However, as discussed in Sect. 8.5, an arc with a relative high arc voltage drop can be present in the cable at the fault location. This effect is included in Case Study II by adding a resistive series fault impedance in the fault loop at the fault location ($R_f = 2.5\Omega$)

Table 10.7 Case study definition. The fault angle is given relative to the voltage at the fault location

Case	Measuring location 1	Measuring location 2	Fault location	Fault resistance (Ω)	Fault angle
I	Joint 0	Joint 33	Joint 27	0	279°
II	Joint 0	Joint 33	Joint 27	2.5	266°
III	Joint 0	Joint 27	Joint 27	0	107°
IV	Joint 27	Joint 33	Joint 27	0	268°

Table 10.8 Expected theoretical and measured arrival instance of first and second coaxial wave from fault location

	$\tau_{CG1,0}$	$\tau_{CG1,33}$	$\Delta\tau_{33-0}$
Theoretical	178.25	39.19	139.06
Measured	181.1	39.84	141.26

and measuring at the same locations as for Case Study I. Case Study III is a study of an ideal fault occurring at Joint 27 with the signals measured at the same location and at Joint 0 (31.4 km cable). In this Case Study, the fault location is at 0 % or 0 km. Case Study IV studies a fault at Joint 27 with the signals measured at the same location and at Joint 33 (6.9 km cable). The two latter cases can help confirm whether a constant coaxial modal velocity can be assumed for the entire cable.

The modal wave velocity is an input parameter for the two-terminal method. Impulse measurements were carried out to determine the modal wave velocities and were presented in Sect. 10.4. These measurements were described in Sect. 10.4, and a mean coaxal wave group velocity of $v_{CG,m} = 1.7333 \cdot 10^8$ m/s was determined. Using both the calculated and the measured coaxial velocities, the expected arrival instance of the fist wave from the fault location at Joint 0 and 33 in case of a fault occurring at $t = 0$ at Joint 27 is calculated and the results are presented in Table 10.8.

Because the instance of fault is unknown when performing the measurements, only the time differences between the arrival instances can be determined. For Case Study I and II, the time difference $\Delta\tau_{33-0}$ shown in Table 10.8 will give the correct fault location. Because the measuring location is at Joint 27 for Case Study III and IV, the time difference $\Delta\tau_{0-27}$ is equal to $\tau_{CG1,0}$ and $\Delta\tau_{33-27}$ is equal to $\tau_{CG1,33}$.

Case Study I

Figure 10.17 shows the fault transient in phase A voltage at Joint 0 and 33 for Case Study I. The fault wave is first observed at Joint 33 whereafter it appears at joint 0.

Figure 10.17a shows that the voltage at Joint 33 will drop towards zero due to the permanent short. The steady state voltage at Joint 0 will decrease due to the change in phase impedance.

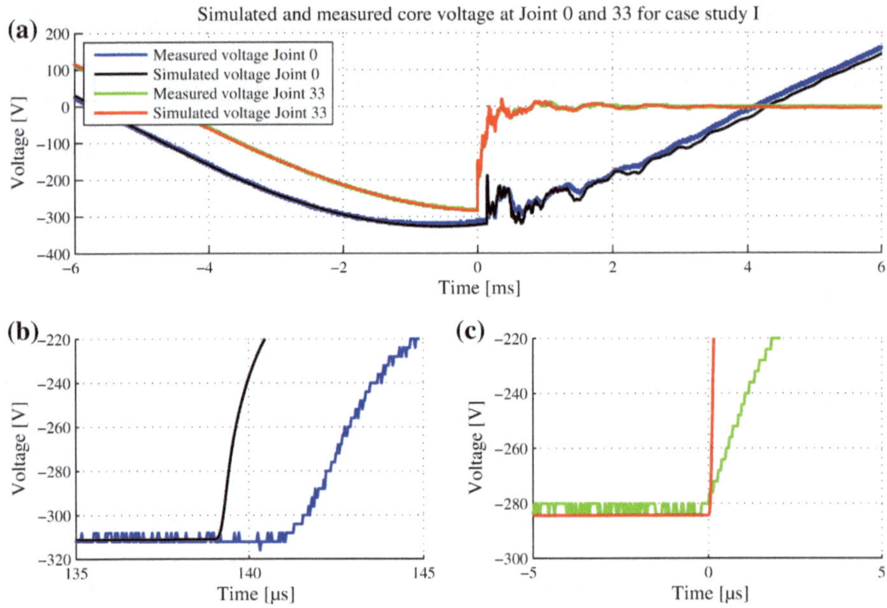

Fig. 10.17 Simulated and measured phase A voltages at Joint 0 and 33 for Case study I

Figure 10.17b and c show that the exact arrival instances of the first coaxial wave group at both joints can be identified within a few samples precision based on a simple visual inspection. As the coaxial velocity used in the simulation is faster than the measured velocity, the simulated fault waves arrive first at Joint 33. Using the inputs found from the visual inspection of the measured signals in Eq. 2.2, the fault location is determined as 31.385 km; an error of −30 m.

Figure 10.17b and c show that the measured voltage collapses slower than predicted by the simulations both at Joint 0 and 33. An ideal breaker is used in the simulations where the Siemens breaker and the wires used to connect it to the cable system will introduce a small series inductance at the fault location. To examine the effect of the additional series inductance in the measurements, the voltage of phase A (faulted phase) at the fault location is shown in Fig. 10.18 at the instance the breaker is closed. The red trace is explained later.

The setup at the fault location is semantically shown in Fig. 10.19a and its electrical equivalent circuit is presented in Fig. 10.19b.

The initial fault voltage wave imposed between core and sheath can be seen as a voltage step applied over the coaxial surge impedances Z_c for each cable. After the breaker is closed, the fault current builds up through the series inductance of the breaker and wires with a time constant determined by the inductance and the surge impedance of the cable. The resulting fault wave voltage imposed at the fault location therefore builds up with the same rate as the current because Z_c is real at

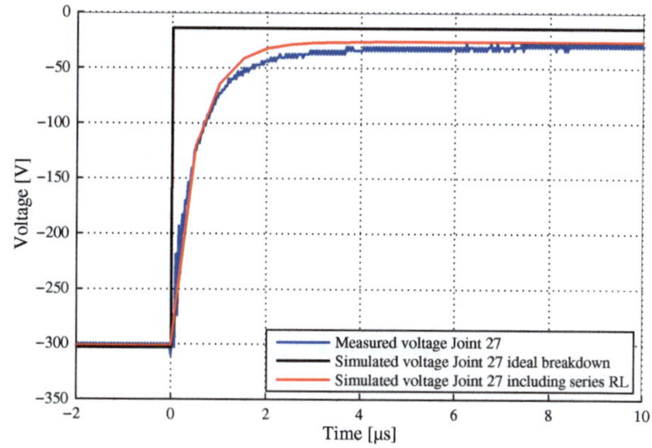

Fig. 10.18 Core voltage at Joint 27 at the instance of fault

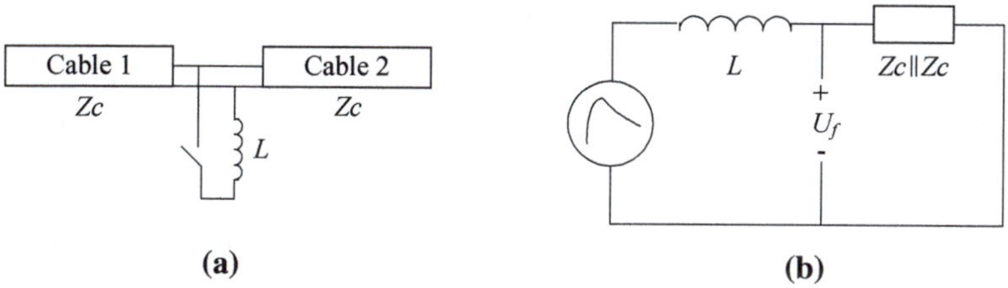

Fig. 10.19 **a** Sketch of breaker at fault location including series inductance and **b** equivalent electrical circuit

high frequencies. The series inductance influences, thus, both the voltage and current waves at the fault location which explains the behaviour at Joint 0 and 33.

A 8 μH inductor (8 m wire) and a 0.3 Ω resistor connected in series are used to model the impedances of the breaker and connecting wires. The effect on the fault location voltages is shown in Fig. 10.18 using the red trace.

The additional inductance introduced by the breaker and the connecting wires is not present in case of a real-life cable fault. Therefore, the elements are not included in the PSCAD/EMTDC model. Especially, additional inductance in the fault loop at the fault location will make exact arrival instance determination at the FLTs harder as fault waves in general contains lower frequency components. Furthermore, the tendency for the fault wave to travel slower due to the lower frequency content is higher. The obtainable accuracy in the measurements is therefore lower than for an ideal voltage collapse cause by insulation breakdown. Because the measured fault signals contain lower frequency components than what is expected in real-life, fault location results using both simulated and measured signals are presented for all case studies. The core current signals of phase A are seen at Joint 0 and 33 in Fig. 10.20.

At Joint 33 (Fig. 10.20c), the simulated and measured core currents are quite different even if the series RL-elements are included at the fault location (not shown). In Sect. 10.4, it has been shown that the core currents are well predicted by the PSCAD/EMTDC model if termination resistors are connected at the cable ends.

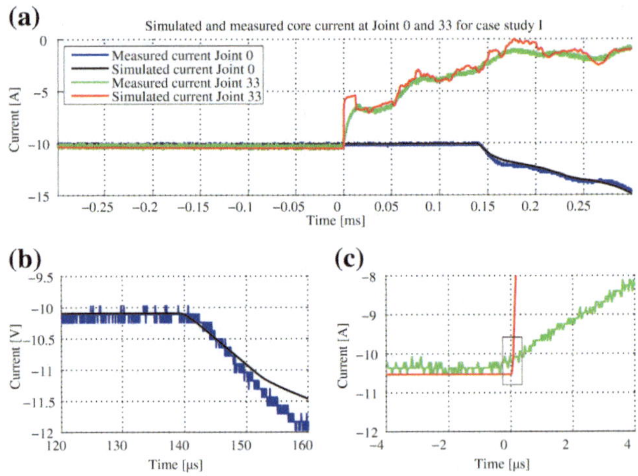

Fig. 10.20 Simulated and measured phase A core current at Joint 0 and 33 for Case Study I

Because of this, it is indicated that the cause of deviation lies in the modelling of the load resistor.

The load resistor connected at Joint 27 is modelled in PSCAD/EMTDC as a real resistor, but in real life, it will have some inductive behaviour due to its coil-like construction. The current cannot instantaneously build up through the windings, thus making the current sloop less steep compared to simulations. Including a small series inductance in the simulations prevents the current from rising fast, and the simulation and measured results become much more comparable. As the load resistor is representing a reflection-less cable line, the exact modelling of this component is not importance for these studies so no attempt is made to obtain a better match between simulation and measurements.

The core and sheath currents at Joint 0 cannot be chosen as input for a fault locator system due to the use of only a transformer at this location. Instead, a time reference for Joint 0 is extracted from the core voltage signal. As all measurements are time synchronised, the use of input signals of different types at each end is not a problem and is also possible when the fault locator system has to be installed in a substation.

Due to the low-frequency content in the current fault waves at Joint 33 (Fig. 10.20c), it is hard to select a unique arrival instance. Instead, a better solution is to select an interval where it seems likely that the fault wave has arrived. Such an interval is defined based on a visual inspection and marked in Fig. 10.20c. Using the core voltage signal to define the arrival instance at Joint 0, and a vector containing the instances defined in the interval shown in Fig. 10.20 (c), Eq. 2.2 will not provide a fault location, but an interval in which it seems most likely that the fault is present. Equation 2.2 is re-written to take a constant and a vector as input:

$$\underline{x} = \frac{l + v_n \cdot (\tau_{CG1,0} - \underline{\tau}_{CG1,33})}{2} \tag{10.2}$$

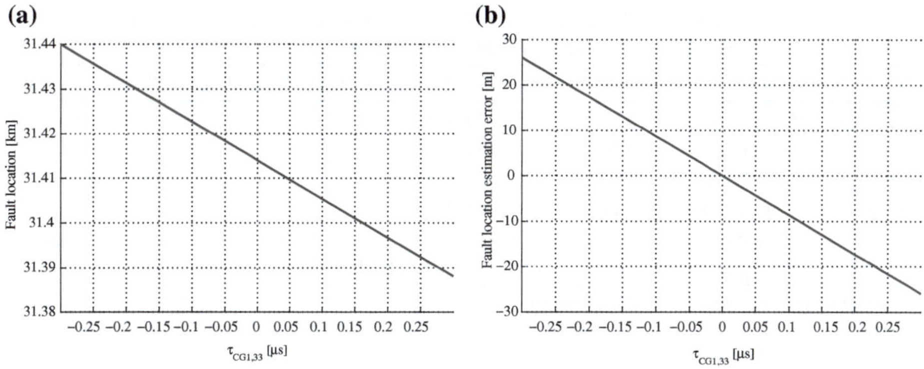

Fig. 10.21 **a** Fault location estimated using intervals defined at Joint 0 and 33 and **b** the resulting error

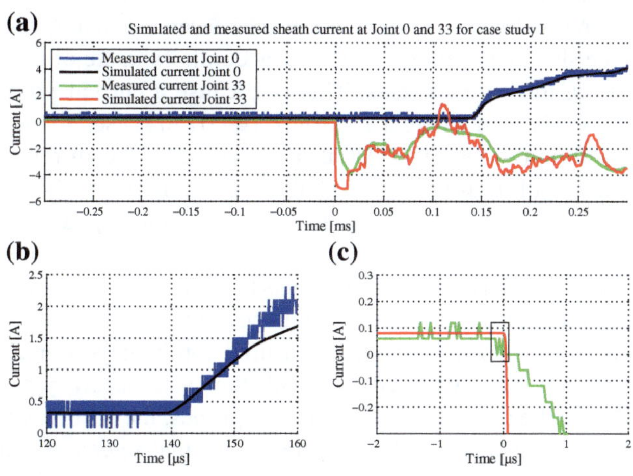

Fig. 10.22 Simulated and measured phase A sheath current at Joint 0 and 33 for Case Study I

where $\underline{x}$ is a vector containing the possible fault locations when the time instance $\tau_{CG1,0}$ and vector $\underline{\tau}_{CG1,33}$ are given as input.

The interval defined in Fig. 10.20c is used in Eq. 10.2 and the resulting estimation of the fault location is shown in Figure 10.21a. The estimated fault location is shown in Fig. 10.21a and the resulting error shown in Fig. 10.21b.

The interval ranges from 31.388 to 31.440 km and is thereby 52 m long. The error compared to the real fault location is between −26 and 26 m thereby placing the actual fault location right in the middle of the interval.

In a real-life fault location situation, it would be beneficial to examine whether the interval determined using vectors as input contains a joint. The possibility of a fault in a joint is higher than the chance for an internal fault to occur in the cable. The sheath currets in phase A at Joint 0 and 33 are presented in Fig. 10.22.

Table 10.9 Fault location estimation for Case Study I using the measured and simulated core voltage, core current and sheath current as input for the two-terminal fault location method

		τ_{0-33} [μs]	Fault location [km]	Interval length [m]	Error [m]
Simulated	Core voltage	139.01	31.145	–	30
	Core current*	139.01	31.145	–	30
	Sheath current*	139.01	31.145	–	30
Measured	Core voltage	141.06	31.385	–	−30
	Core current*	[141.1;141.7]	[31.388;31.440]	52	[−26;26]
	Sheath current*	[141.3;141.8]	[31.409;31.447]	37	[−5;33]

* the core voltage signal is used to determine τ_0

As seen from Fig. 10.22c, that an interval with advantage can be selected for the sheath current as well. Using the intervals shown in Fig. 10.22c, the fault location is estimated to be between 31.409 to 31.447 km and the interval is thereby 37 m long. The error compared to the real fault location is between −5 and 33 m.

The use of the sheath current gives a more accurate estimate of the fault location compared to using the core currents. This is because a current probe of the type Tektronix TCP0030 is used where a Tektronix TCP0150 is used to measure the core current. The resolution of the TCP0030 is more suited for the current which it measures and the transient is thus more identifiable (see Table 10.1). The results for Case Study I using the two-terminal method and different input signals are summarised in Table 10.9.

Table 10.9 shows good results for all signals types when the simulated signals are used. The measured signals have lower frequency components due to the inductive elements at the fault location. Using the measured core voltage signals as input, the fault location is, however, still predicted with the same accuracy as found in the simulations. Due to the resolution of the current probes, the accuracy is better when the sheath current signals are analysed. This shows that selecting a suitable range and if possible a data recording unit with better vertical resolution is important.

Case Study II

Figure 10.23 shows the phase A voltage at Joint 0 and 33 for Case Study II where a 2.5 Ω fault resistor is connected in the fault loop at Joint 27.

The reason for the slow collapsing voltages was found in the extra inductance introduced by the fault resistance in the fault loop at the fault location. For Case Study II, the series resistor is connected in the fault loop to represent an arc voltage drop. This resistor is constructed using a coil-like winding method and is therefore an additional series inductance much larger than the inductance introduced by the breaker and connecting wires. With the same argumentation as made for Case Study I, this effect is not included in the simulations.

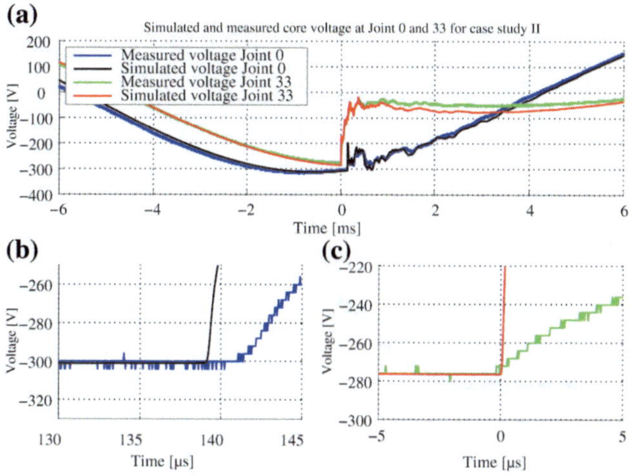

Fig. 10.23 Simulated and measured phase A voltages at Joint 0 and 33 for Case Study II

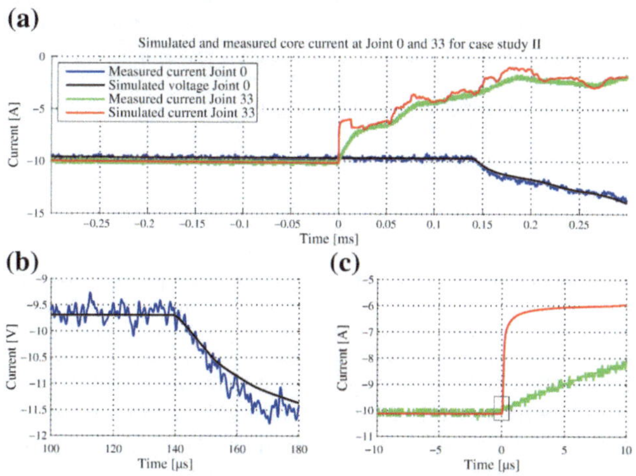

Fig. 10.24 Simulated and measured phase A core current at Joint 0 and 33 for Case Study II

Using the measured core voltages, the fault location is estimated as 31.414 km; an error of 1 m. The almost nonexisting error is not an indication of a better fault location estimation. With the used sampling frequency (31.25 MHz), the highest obtainable accuracy is 3 m. The lengths of the cable sections are provided by cable manufactures and can vary a bit. The results show, however that an arc resistance of a realistic value will not lower the accuracy of the fault location if the voltage waves are used as input. The core currets in phase A at Joint 0 and 33 are presented in Fig. 10.24.

As for Case Study I, an interval in which the fault wave has most likely arrived is defined for the core current at Joint 33. The fault location is estimated between

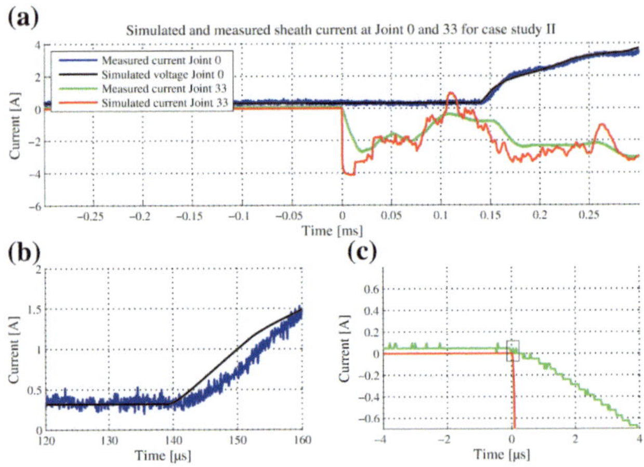

Fig. 10.25 Simulated and measured phase A sheath current at Joint 0 and 33 for Case Study II

Table 10.10 Fault location estimation for Case Study I error using the measured and simulated core voltage, core current and sheath current as input for the two-terminal fault location method

		τ_{0-33} [μs]	Fault location [km]	Interval length [m]	Error [m]
Simulated	Core voltage	139.00	31.442	–	27
	Core current*	139.00	31.442	–	27
	Sheath current*	139.00	31.442	–	27
Measured	Core voltage	141.39	31.414	–	1
	Core current*	[141.0;141.8]	[31.379;31.449]	69	[−34;35]
	Sheath current*	[139.6;143.9]	[31.380;31.417]	37	[−34;3]

For the signals marked with '*', the core voltage at Joint 0 is used as the Joint 0 time reference

31.379 and 31.449 km. The error is between −34 and 35 m which makes the interval length 69 m. The sheath current in phase A at Joint 0 and 33 are shown in Fig. 10.25.

Using the sheath currents as input with the interval defied in Fig. 10.25b and c, the fault location is estimated between 31.380 and 31.417 km. The error is between −34 and 3 m. The interval length is 37 m. All results for Case Study II are summarised in Table 10.10.

A very accurate fault location can still be determined using the core voltage signals even with a fault resistor (extra inductance) included in the fault loop. The use of the sheath current gives again the best estimation of the fault location when using current signals as the probe resolution is highest.

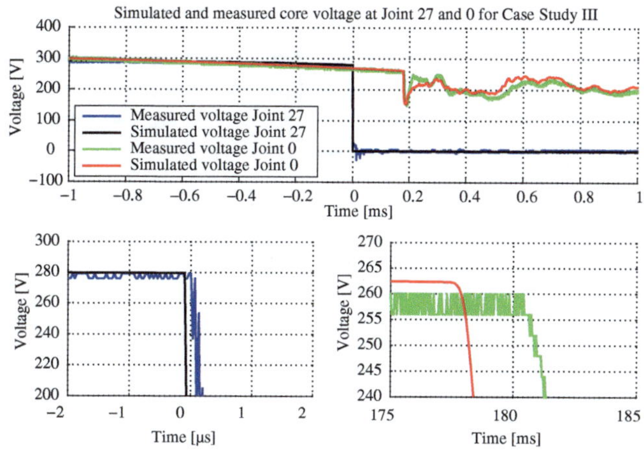

Fig. 10.26 Simulated and measured phase A voltages at Joint 27 and 0 for Case study III

Case Study III and IV

The results of Case Study III and IV are presented together. For these two case studies only the core voltage is analysed. The same PSCAD/EMTDC simulation model is used, but only the sections from either Joint 0 to 27 or from 27 to 33 is included.

A fault is applied at Joint 27 and the transient response measured at the same location as well as at Joint 0 is shown in Fig. 10.26.

Figure 10.26 shows that the arrival instance can be determined exactly at both Joints. At Joint 27, where the fault is applied, a voltage collapse with some jitter is seen. This is due to a non-ideal closing of the breaker. However, the first wave is clearly identified. Using the measured arrival instances at Joint 27 and 0, the fault location is determined as 44 m; an error of 44 m as the fault location is directly at Joint 27. Using the simulated signals as input gives a fault location of 42 m.

In Case Study IV, the fault is applied at Joint 27 and the response measured at this location and at Joint 33. The core voltages at these two locations are shown in Fig. 10.27.

Using the measured voltages, the fault location is estimated as 42 m seen relative to Joint 27 where the fault is applied. Using simulated voltages as input, the fault location is determined with an error of 23 m. The results of both Case Study III and IV show that a constant coaxial wave velocity determined on the full system can be used for any fault location on the same system with little error.

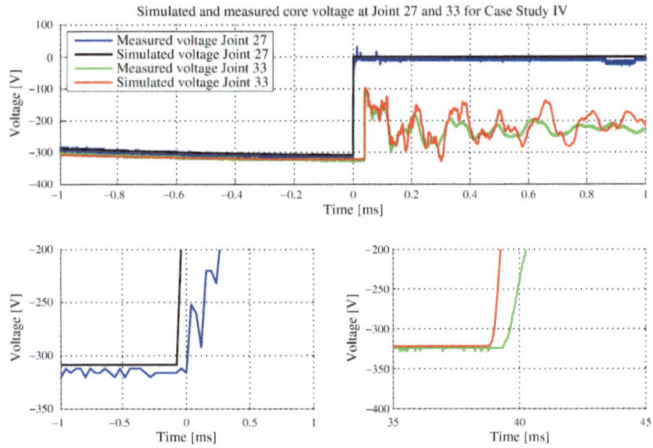

Fig. 10.27 Simulated and measured phase A voltages at Joint 27 and 33 for Case study IV

10.5.1 Summary

Field measurements showing that it is possible to use a travelling wave-based fault location method on the Anholt land cable have been carried out. Using the two-terminal method with core voltages, core currents or sheath currents as input, accurate fault location is possible. The initial fault wave created during the field measurement does not resemble an ideal voltage collapse well due to additional inductance at the fault location. However, it is still possible to obtain high accuracy when estimating the fault location.

A method to deal with low fault wave frequency content was presented. The arrival instance in the core and sheath current signals was described by a suitable interval instead of a single unique value. The result is an interval in which the fault is localised. In case of using both the core and sheath currents with the proposed method, the fault location is found within the intervals.

Two assumptions were made when the two-terminal method was proposed.

- A constant coaxial wave velocity can be used as input to the fault locator system.
- The attenuation of the coaxial waves will not compromise the ability to locate faults.

Both assumptions hold true in case of the field measurements conducted on the Anholt cable. As the behaviour of the coaxial waves predicted by PSCAD/EMTDC show the same degree of damping as found in the field measurements, it also holds true, that the simulations presented earlier in this thesis and the conclusions drawn from these can be trusted. Therefore:

Fault location using a synchronised two-terminal method is applicable on cross-bonded cables with use of the coaxial wave velocity and fault signals analysed directly in the time domain.

The conclusions made based on the Anholt measurements are also valid for fault location where other single-core cables are used as the coaxial waves are only dependent on a few materials found in all high voltage coaxial cables.

References

1. U.S. Gudmundsdottir, *Modeling of long High Voltage AC cables in Transmission Systems*, 2nd edn. (Energinet.dk, Denmark, 2010). ISBN 978-87-90707-73-6
2. Stanford Research Systems. PRS10 Rubidium frequency standard with low phase noise. 2012
3. A. Morched, B. Gustavsen, M. Tartibi, A universal model for accurate calculation of electromagnetic transients on overhead lines and underground cables. Power Delivery IEEE Trans. **14**(3), 1032–1038 (1999)
4. B. Gustavsen, J. Nordstrom, Pole identification for the universal line model based on trace fitting. Power Delivery IEEE Trans. **23**(1), 472–479 (2008)
5. B. Gustavsen, Time delay identification for transmission line modeling. in *Signal Propagation on Interconnects, 2004. Proceedings. 8th IEEE Workshop on*, 2004, pp. 103–106
6. B. Gustavsen, A. Semlyen, Simulation of transmission line transients using vector fitting and modal decomposition. Power Delivery IEEE Trans. **13**(2), 605–614 (1998)
7. A. Ametani, A general formulation of impedance and admittance of cables. Power Apparatus Syst. IEEE Trans. **PAS-99**(3), 902–910 (1980)
8. B. Gustavsen, J.A. Martinez, D. Durbak, Parameter determination for modeling system transients-part ii: Insulated cables. Power Delivery IEEE Trans. **20**(3), 2045–2050 (2005)
9. C.F. Jensen, F. Faria da Silva, C.L. Bak, and W. Wiechowski. Switching studies for the horns rev 2 wind farm main cable. in *International Conference on Power Systems Transients (IPST)*, 2011
10. W.L. Weeks and Y.M. Diao, Wave propagation characteristics in underground power cable. Power Apparatus Syst. IEEE Trans. **PAS-103**(10), 2816–2826 (1984)
11. B. Gustavsen, Panel session on data for modeling system transients insulated cables. in *Power Engineering Society Winter Meeting, 2001*, vol. 2 (IEEE, 2001), pp. 718–723

Chapter 11
The Wavelet Transform and Fault Location on Crossbonded Cable Systems

Over the last years, the Wavelet transform (WLT) has gained a lot of attention for solving fault location problems on transmission lines [1–7]. A fault locator system which do not require skilled personal to operate it is preferable as fault on transmission lines are seldom and the skill set of the maintenance crew needs to be keep up to date. The Wavelet Transform has the ability to determine the fault location automatic based on signals recorded at the fault locator terminals and thereby eliminate the need for skilled personal.

The Wavelet transforms offers the ability for multiresolution analysis. The transform provides both high time and frequency resolution at the same time [8]. One of the major shortcomings for use of older analysing techniques as for instance the short Fourier Transform is that a window of constant size is used. This poses a problem for the analysis of transient signals which can have wide frequency spectrums and are non-stationary in time. According to the uncertainty principle, a higher time resolution will result in lower frequency resolution and vice versa [9]. The wavelet transform overcomes this problem by allowing variable window sizes so that long time intervals are used at low frequencies and short time intervals are used at high frequencies. This is illustrated in Fig. 11.1.

The WLT decomposes the time domain signal into so-called scales using dilated and contracted versions of a 'mother wavelet'. The low scales contains the high frequencies and the high scales contain the low frequencies.

Wavelets are functions which are oscillatory of nature, they must decay quickly to zero, and have an average value of zero. The wavelet coefficients (results from the transform) can be seen as a measure of the similarity between the wavelet and the analysed signal. For an ideal non-faulted voltage or current signal, the fundamental will be dominating and the scales that contain the high frequency components will have small wavelet coefficients (WLC). When the fault wave arrives at the fault location terminal, the WLCs at the same scales will increase in value. For fault location purposes, the calculated WLCs at one or more pre-chosen scales are compared to a threshold value. If the value of the WLCs exceeds this value, a fault has occurred,

C. F. Jensen, *Online Location of Faults on AC Cables in Underground Transmission Systems*, Springer Theses, DOI: 10.1007/978-3-319-05398-1_11,

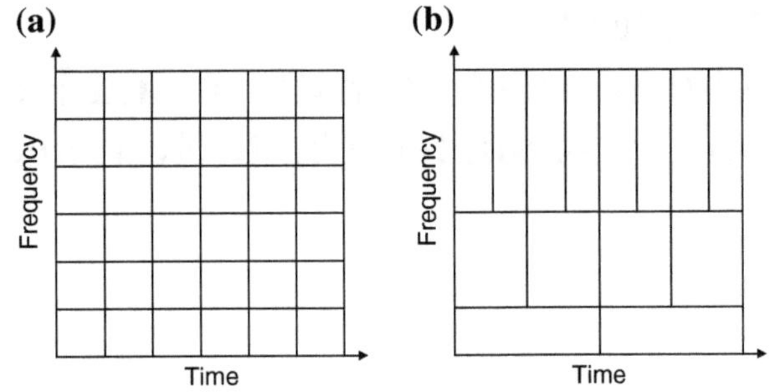

Fig. 11.1 **a** Constant size windows and **b** variable size windows

and because the WLCs are localised in time, they can be used to calculate the fault location.

The choice of the 'best suited' mother wavelet is well discussed in literature [8, 10, 11]. Because the WLCs are a measurement of the best fit between the wavelet at a given scale and the analysed fault signal, different faults on different systems will have different 'optimal' mother wavelets. The choice of mother wavelet is, therefore, often based on an engineering evaluation for the given system. Popular mother wavelets used for fault location purposes are members of the Dubieties (Db) family, the Morlet, the Symlet and the Haar wavelet. Furthermore, research has been published where unique wavelets are designed for the specific purpose of fault location either off- or online [3, 12].

11.1 The Wavelet Transform

The continuous wavelet transform as functions of scale and position (time) taken on the signal $s(t)$ is calculated as shown in Eq. 11.1.

$$C(a, b) = \int_{-\infty}^{\infty} s(t) \frac{1}{\sqrt{a}} \psi^* \left(\frac{t - b}{a} \right) dt \tag{11.1}$$

where ψ^* denotes the complex conjugate of the daughter wavelet which is a time shifted and scaled version of the mother wavelet $\psi(t)$, b is the time shifting factor and a is the scale factor which is inversely proportional to the frequency. The numerical implementation of Eq. 11.1 looks as follows:

$$C(a, iT_s) = T_s \frac{1}{\sqrt{|a|}} \sum_{n=0}^{N-1} \left[\psi^* \left(\frac{nT_s - iT_s}{a} \right) s(nT_s) \right] \tag{11.2}$$

where T_s is the sampling time, N is the total number of samples, i is an integer representing the number of sampling times the daughter wavelet is shifted in time.

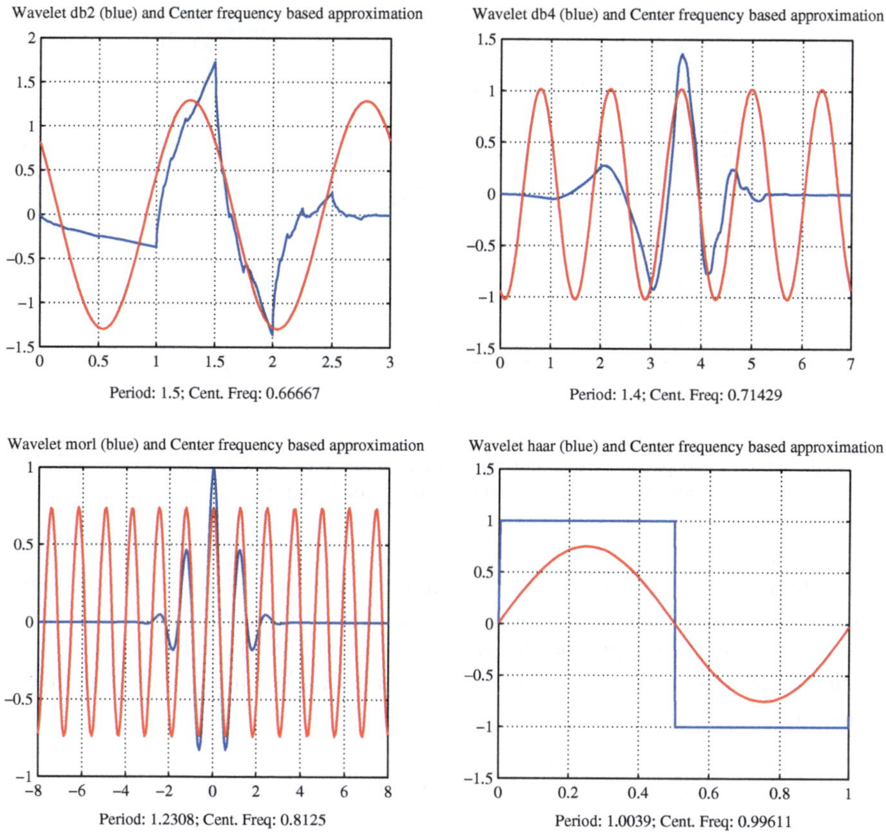

Fig. 11.2 Center frequency of **a** the Daubechies 2, **b** the Daubechies 4, **c** the Morlet and **d** the Haar wavelet

11.1.1 Scale and Frequency

Scales are used to represent energy in the analysed signal at different "frequency bands". For fault location problems, a scale (frequency band) needs to be selected in where the analysis of the faulted transient signals are performed. Different fault signals will have different frequency content and will therefore be found in different scales. The Wavelet Transform can thus provide a measure of frequency content of faulted signals localised in time. The concepts of center frequency and pseudo-frequencies are introduced in order to relate the mother wavelet and the different scales to a certain frequency.

The frequency at which the modulus of the Fourier components of the mother wavelet is maximised is defined as the center frequency. Therefore, the centre frequency captures the dominant oscillation of the wavelet and can be used to define a frequent for any scale. In Fig. 11.2 it is seen how the center frequency is determined for four wavelets commonly used for fault location purposes.

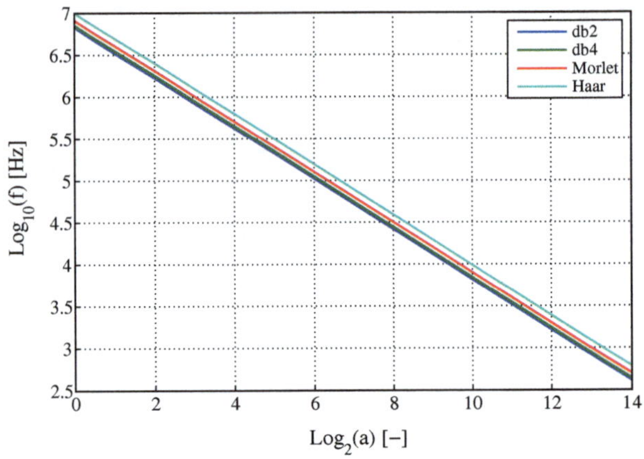

Fig. 11.3 Scale versus pseudo-frequency calculated using the center frequency of the Db2, Db4, Morlet and Haar mother wavelet at a sampling frequency of 10 MHz

The Fourier transform identifies the frequency components of a periodic signal and maps them in the frequency domain where, as mentioned, the Wavelet transform uses scales [13]. The Continuous Wavelet Transform (CWT) can be interpreted as a frequency based filtering of the signal by rewriting the CWT in Eq. 11.1 as an inverse Fourier Transform:

$$C(a, b) = \frac{1}{2\pi} \int_{-\infty}^{\infty} \hat{s}(\omega)\sqrt{a}(\hat{\psi}(a\omega))^{*} e^{j\omega b} d\omega \tag{11.3}$$

where $\hat{s}(\omega)$ and $\hat{\psi}(\omega)$ are the Fourier transforms of the signal $s(t)$ and the Wavelet ψ.

Equation 11.3 shows that for increasing scale, the support in the frequency domain is reduced. Furthermore, Eq. 11.3 shows that the so-called pseudo-frequency of $\psi(a\omega)$ is equal to f_C/a where f_C is the center frequency of the mother wavelet. Different scales will thus contain different frequency bands of different bandwidth determined by the band-pass features of the daughter wavelet used at the given scale. The pseudo-frequency for the scale a is calculated based on the center frequency as shown in Eq. 11.4.

$$F_a = \frac{f_c}{aT_s} \tag{11.4}$$

where T_s is the sampling period and f_c is the center frequency.

The scale/frequency relationships of the Dubieties 2 (Db2), Dubieties(Db4), Morlet and Haar wavelet are seen in Fig. 11.3.

For instance, for determining the pseudo-frequency for the Haar wavelet at scale 4 with a sampling frequency of 10 MHz, the center frequency of 0.99611 is divided by the scale number multiplied the sampling time:

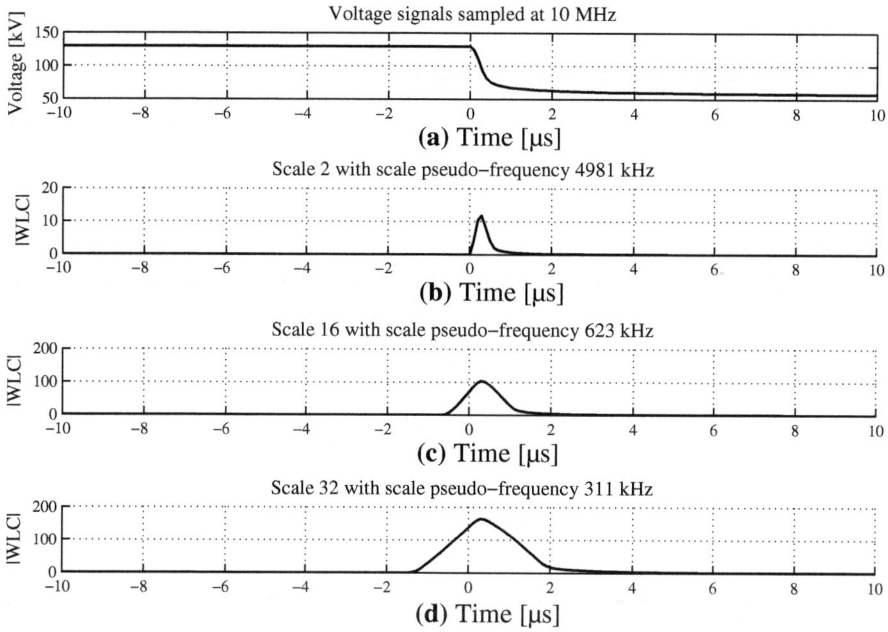

Fig. 11.4 **a** Time domain Terminal A voltage and the Haar wavelet coefficients calculated at scale **b** 2, **c** scale 8 and **d** scale 16

$$F_a = \frac{0.99611}{4 \cdot 1 \cdot 10^{-7}} = 2.4903 \text{ MHz} \tag{11.5}$$

11.1.2 The Wavelet Transform for Detection of Singularity

When a fault wave arrives at a substation's terminals, a singularity in the signal is seen. The CWT can detect both the presents of a singularity and when it occurs in time by identifying frequency content of a different nature than under normal conditions. The low scales will as mentioned contain the highest frequency components and the higher the scales, the lower the frequency components. It is of interest to identify which frequencies are found in fault signals on both short and long crossbonded cables and how these components are related to the fault- and system parameters. Furthermore, the identification of the best suited mother wavelet for the analysis is necessary.

Choice of Scale

Several considerations must be made before choosing the most optimal scale for fault location purposes. Figure 11.4 shows the normalised Haar WLCs calculated at scale

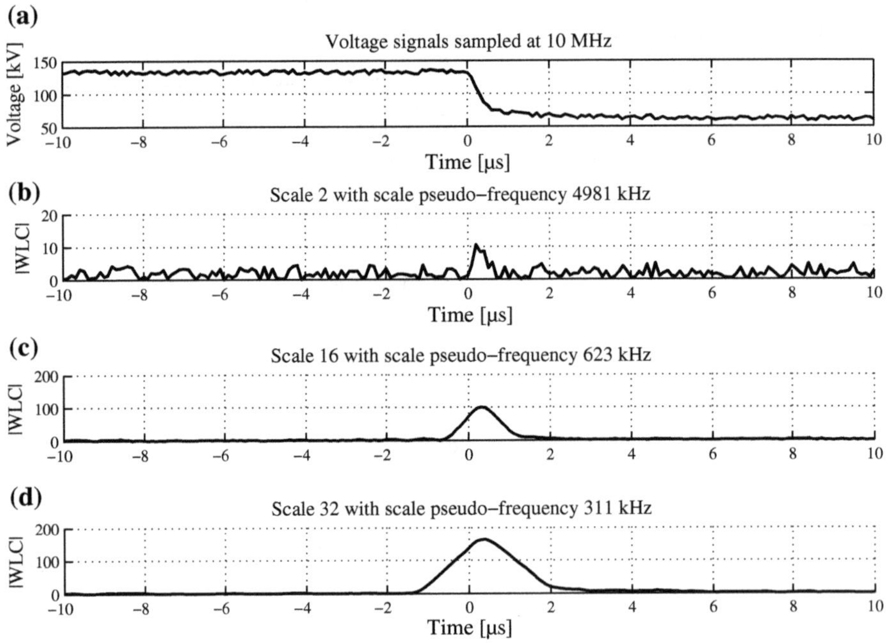

Fig. 11.5 **a** Time domain Terminal A voltage added 2 % white noise and the Haar wavelet coefficients calculated at **b** scale 2, **c** scale 8 and **d** scale 16

2, 16 and 32 for the Terminal A voltage on a 60 km crossbonded cable with the fault location 6 km from Terminal A. The time domain signal is shifted so the fault wave arrives at the terminal at $t = 0$.

By comparing Fig. 11.4b–d it is seen that the finer scales enables the most accurate determination of the arrival instance of the fault waves and therefore the most accurate estimation of the fault location. The WLCs are, however, calculated based on simulated data. Real-life signals do contain some high frequency (HF) noise and this will affect the low scales the most. Figure 11.5 shows the WLCs of the same signal as presented in Fig. 11.4, but added 2 % white noise.

Figure 11.5 shows why the high scales can be difficult to use for fault location. HF signal noise is dominant at the finer scales that at the same time offers the best ability to determined the arrival instance of the fault wave. The choice of scale becomes therefore a compromise between accuracy and the ability to pick up the wave at all. This is an obvious problem because as the scale number increases, the accuracy of the fault location decreases and as the fault location is not known it can be difficult to predict the frequency content of the arriving fault waves.

11.2 Automatic Fault Location on Crossbonded Cables Using the Wavelet Transform

The ability to automatically determine the location of a fault without the need of trained personnel is desirable. However, acting on faulted information caused by malfunction of a fault locator system can be costly and time consuming. The Wavelet Transform's ability to determine the fault location automatically on a few selected cable systems is examined in this section. First, however, a strategy for detecting the fault wave arrival instance is discussed.

11.2.1 Automatic Fault Location Strategy

A common way to determine if a fault wave has arrived using the WLT is to examine the WLC at a given scale and compare them to a pre-determined threshold value. Which scale is best suited will depend on the frequency content and the noise level in the analysed signal. Using low scales, a high accuracy can be obtained where the higher scales determine the arrival instance less precisely, but are more noise resilient. For a short cable system, high frequency content is expected in the fault signals whereas on longer cables, the frequency content will be lower. Furthermore, the different mother wavelets will have different abilities to capture the incoming fault waves depending on the characteristics of the particular wave. All this indicates that a unique best suited mother wavelet and scale can be found for each system configuration, fault location and fault type. As the fault location and thereby the fault signal frequency content is unknown in a real-life fault location situation where an automatic system is desired, a 'one size fits all' solution must, however, be chosen for every system if no adaptive methods are utilised.

The threshold value can either be set in advance or calculated online. To reduce the influence of the noise in the coefficients the squares of the wavelet coefficients can be calculated [14]. A way to calculated it online is to find a value for a given scale based on a certain number of WLC calculated pre-fault. Commonly, the mean value of these WLC is multiplied a safety margin value to make the triggering immune for noise [14].

11.2.2 Case Studies

In the following section, fault location using wavelets is examined. Four cases are chosen, widely representing the cable lines installed in the future Danish grid. These are:

1. The measured Anholt data of Case Study I described in Sect. 10.5.
2. A 10 km crossbonded cable system with 2 major sections and one additional cable connected at each the FLT.

3. A 20 km crossbonded cable system with 2 major sections and one additional cable connected at each the FLT.
4. A 60 km crossbonded cable system with 10 majors and one additional cable connected at each the FLT.

Case Study 1 is the Anholt field measurements presented in Sect. 10.5.

Case Study 2, that studies fault location on a 10 km cable system with one additional cable connected at each substation is included as it represents half of the cables planned in the future Danish transmission system (see Fig. 8.1b). In this case, a single phase A fault is applied 3.33 km from Terminal A in the first crossbonding. Using the same substation configuration as for Case Study 2 and increasing the length of the crossbonded cable to 20 km, Case Study 3 covers 77 % of all Danish lines. In this case, a single phase A fault is applied 3.33 km from Terminal A as well.

Case Study 4, with a 60 km cable and one additional cable at each terminal will for a fault close to Terminal A (6 km) represent a worst case scenario for all Danish cables. At Terminal A, high frequency components are expected where at Terminal B, the frequency content will be lower and the relative magnitude change in the voltage due to the fault wave will be low as well.

The threshold values for all cases are calculated as the mean of the pre-fault 1000 wavelet coefficients multiplied with a safety margin value of 4. All mother wavelets, available in the MATLAB 'Wavelet toolbox' are studied, but only a few of the most interesting are presented here. The chosen mother wavelets are the Haar, Daubechies (Db2), Daubechies (Db4), Morlet and Symlet4. The WLCs are calculated at scale 2, 4, 8, 16, 32, 64 and 128 with the sampling frequency in the simulation set to 10 MHz and the Anholt field data is downsampled three times to make the results at the different scales comparable (10.42 MHz). All WLCs are calculated based on the voltage signals, but as one additional cable is connected at each terminal, the core and sheath currents are equally advantageous to use.

Signal Pre-Conditioning for Optimal Wavelet Performance

It was shown in Sect. 6.2.1, that a part of the fault wave on the faulted conductor is reflected onto the two healthy conductors as the fault wave propagates along the crossbonded cable. A fault wave with smaller magnitude is harder to detect by the Wavelet Transform and it is thereofre proposed in this thesis to pre-condition the signal before it is analysed by the WLT simply by adding the three phase signals recorded at the FLT point by point. This combines all the fault wave energy into one new signal and furthermore, the power frequency signal is canceled. The method can be used for both the core voltages, core currents and sheath currents as the fault wave is reflected positive onto the two additional conductors in all cases (see Sect. 6.2.1).

In Fig. 11.6, the square of the Haar wavelet coefficients at Scale 5 are presented for the Anholt core voltage signals with and without the proposed pre-conditioning. The threshold line (THL) value for each signal is calculated based on 1000 pre-fault WLCs.

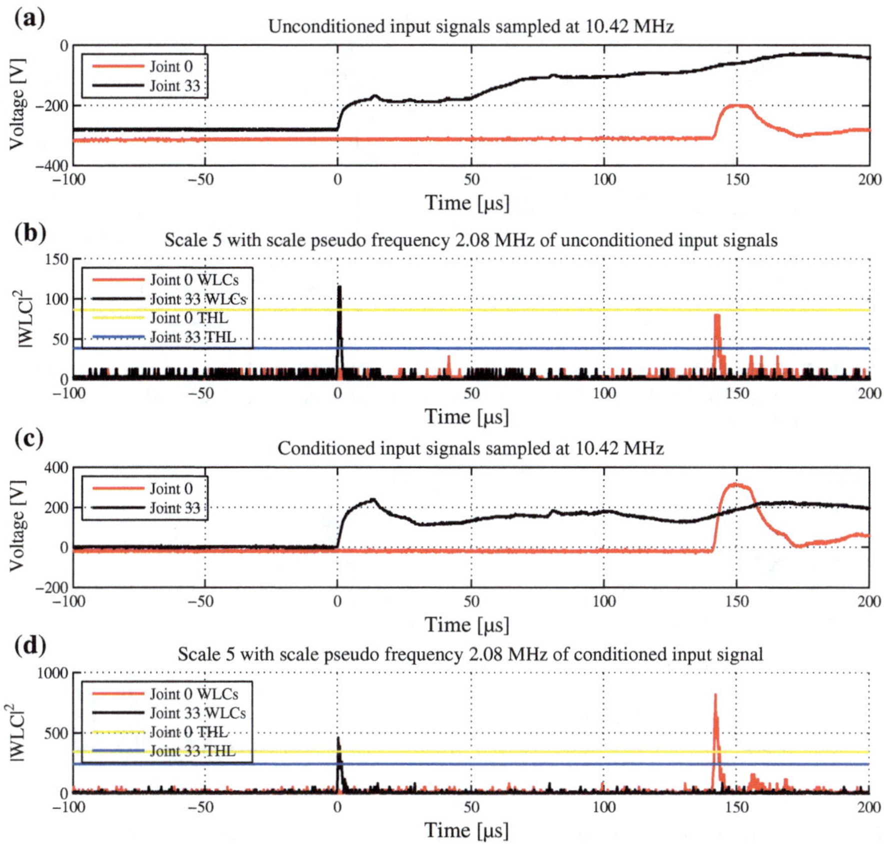

Fig. 11.6 **a** Anholt Case Study I signals, **b** the squares of the Haar WLC at scale 5 with the threshold line (THL), **c** Anholt Case Study I conditioned signals, **d** the squares of the Haar WLC at scale 5 with the threshold line (THL) for the conditioned signal

By comparing Fig. 11.6b–d, it can be seen that by pre-conditioning the signal in the proposed way, the fault wave can be picked up at finer scales. In Fig. 11.6b, which shows the WLCs calculated on the basis of a non-pre-conditioned signal, the wave arrival instance at Joint 33 is detected, but the wave at Joint 0 is not. In Fig. 11.6d, both the wave at Joint 0 and 33 are, however, detected. Using finer scales enables a more precise estimate of the fault location as discussed earlier in this chapter.

Using the signal pre-conditioning method, the fault location estimation error at different scales for selected mother wavelets are presented for the Anholt Case Study I and the 10 km crossbonded cable in Fig. 11.7a and b respectively.

It is seen from Fig. 11.7a, that the Haar wavelet performs well for the Anholt Case Study I. At scale 5, which is the first scale to pick up the wave at Joint 0, the fault location is estimated with an error of 22 m. The Haar wavelet pick up the wave at all other scales included and determines the fault location with good results. The Db2

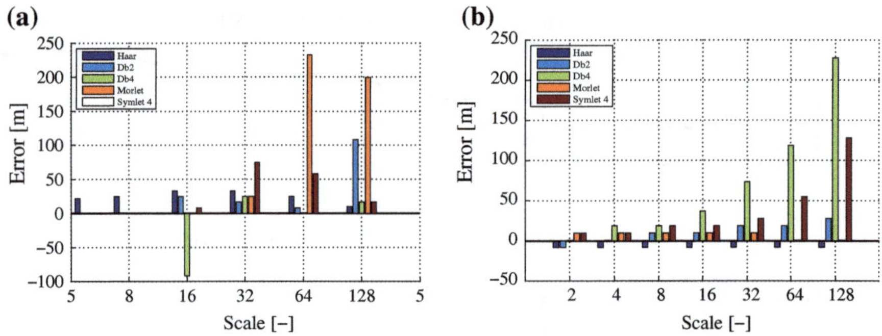

Fig. 11.7 Estimated fault location error for the Haar, Db2, Db4, Morlet and Symlet 4 wavelet at scale 2, 4, 8, 16, 32, 64 and 128 for the Anholt measurements. **a** Case Study I and **b** 10 km crossbonded cable

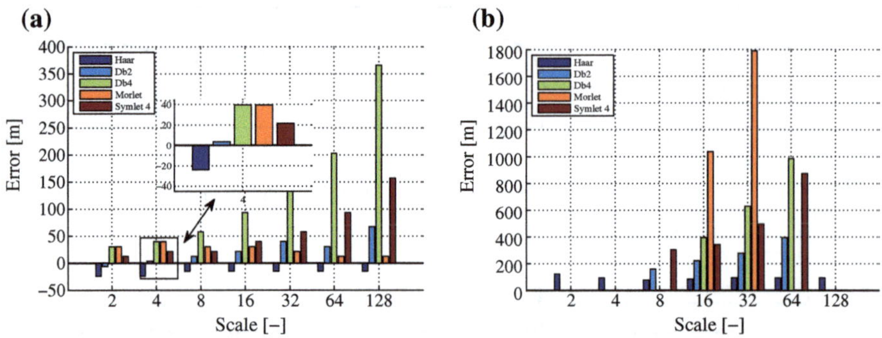

Fig. 11.8 Estimated fault location error for the Haar, Db2, Db4, Morlet and Symlet 4 wavelet at scale 2, 4, 8, 16, 32, 64 and 128 for the **a** 20 km and **b** 60 km crossbonded cables

wavelet performs well too and picks up the arrival instance at Scale 16, 32 and 64 with a maximum error of 34 m. At scale 128 the error is, however, 108 m.

Figure 11.7b shows that most wavelets perform well at all scales. The Haar wavelet picks up the fault waves at all scales with a maximum error of 8 m. As the 10 km system is representative for most lines in the future Danish grid, automatic online fault location is in fact possible on most lines using the proposed method with a clever choice of wavelet and scale.

In Fig. 11.8a and b, the fault location estimation errors on the 20 and 60 km crossbonded cables are shown respectively.

Figure 11.8a shows high accuracy when using the Haar wavelet for automated fault location on the 20 km crossbonded cable. An error between −14 and −23 m is found using all scales. The findings here indicate that automated faut location can be implemented on 77 % of all Danish cable lines.

Figure 11.8b shows that the determination of the fault location using an automated method based on wavelets becomes much harder as the cable length increases. Again, the Haar wavelet performs best with an error at the 5th, 8th and 16th scale of approximately 85 m. If signal pre-conditioning of the signal is not carried out, the waves at both ends are picked up first at the 8th scale with an error of 213 m. Using a visual inspection method the estimation error is, however, reduced to 40 m. This indicated that a visual inspection of the fault signals should always be carried out before any digging is started.

In all cases it is found that the Haar wavelet performs the best. This is most likely because the initial fault wave created at the fault location appears as a square wave superimposed on the power frequency signal and as the Haar Wavelet is a square wave as well (see Fig. 11.2), high correlation is calculated by the Wavelet Transform.

11.2.3 Summary

Automatic fault location on crossbonded cables is possible on most cable systems installed in the future Danish transmission system. As the cable length increases, the ability for the WLT to accurately pick up the incoming fault waves decreases, thereby reducing the accuracy of the fault location.

The Haar wavelet performs best on all studied systems indicating that this particular wavelet is a good choice for analysing voltage waves on cable system. As the performance of the automatic fault locator is very system dependent, especially for longer cable lines, studies for such cases seem preferable before a choice is made to install equipment for a given line. For shorter lines, a 'one fits all' solution is possible.

It was shown in Chap. 8 that a visual signal inspecting method will give good results in most realistic cases. It is therefore recommended in this thesis to develop a fault location system that combines the wavelet- and a visual based inspection method as a best overall solution. A simple visual inspection by a trained person will reduce the chance of acting on faulted automatic calculated fault location information and reduce the error in estimating the fault location.

References

1. M. Paolone, A. Borghetti, C.A. Nucci, An automatic system to locate phase-to-ground faults in medium voltage cable networks based on the wavelet analysis of high-frequency signals. In *PowerTech, 2011 IEEE Trondheim*, (2011), pp. 1–7
2. A. Borghetti, M. Bosetti, C.A. Nucci, M. Paolone, A. Abur, Integrated use of time-frequency wavelet decompositions for fault location in distribution networks: theory and experimental validation. Power Delivery IEEE Trans. **25**(4), 3139–3146 (2010)
3. A. Borghetti, M. Bosetti, M. Di Silvestro, C.A. Nucci, M. Paolone, Continuous-wavelet transform for fault location in distribution power networks: definition of mother wavelets inferred from fault originated transients. Power Systems IEEE Trans. **23**(2), 380–388 (2008)
4. Zhengyou He, Fu Ling, Fault detection and classification in ehv transmission line based on wavelet singular entropy. Power Delivery IEEE Trans. **25**(4), 2156–2163 (2010)

5. S. Lin, Z.Y. He, X.P. Li, Q.Q. Qian, Travelling wave time-frequency characteristic-based fault location method for transmission lines. IET Gener. Transm. Distrib. **6**(8), 764–772 (2012)
6. F.B. Costa, B.A. Souza, N.S.D. Brito, Effects of the fault inception angle in fault-induced transients. IET Gener. Transm. Distrib. **6**(5), 463–471 (2012)
7. M. Korkali, A. Abur, Fault location in meshed power networks using synchronized measurements. In *North American Power Symposium (NAPS) 2010*, pp. 1–6
8. O.M.K.K. Nanayakkara, A.D. Rajapakse, R. Wachal, Location of dc line faults in conventional hvdc systems with segments of cables and overhead lines using terminal measurements. Power Delivery IEEE Trans. **27**(1), 279–288 (2012)
9. A.V. Oppenheim, R.W. Schafer, J.R. Buck, *Discrete-time Signal Processing*, 2nd edn. (Prentice-Hall, New Jersey, 1989)
10. A.M. Elhaffar, Power transmission line fault location based on current traveling waves. Doctoral Dissertation, Helsinki University of Technology Faculty of Electronics, Communications and Automation Department of Electrical Engineering, 2008
11. I. Daubechies, *Ten Lectures on Wavelets* (SIAM, Philadelphia, 1992)
12. P. Argyropoulos, Customized wavelets for fault location in power systems. Northeastern University, Master thesis edition, 2010
13. R.G. Lyons, *Understanding Digital Signal Processing*, 3rd edn. (Carl Hanser Verlag GmbH & Co, Munich, 1987)
14. M. Korkali, H. Lev-Ari, A. Abur, Traveling-wave-based fault-location technique for transmission grids via wide-area synchronized voltage measurements. Power Systems IEEE Trans. **27**(2), 1003–1011 (2012)

Chapter 12
Development of a Fault Locator System for Crossbonded Cables

After a study of the different fault location techniques, it is clear that the online travelling wave method is well suited for crossbonded cables. It was shown in Chaps. 7 and 10 that coaxial signals analysed in the time domain will provide a good input for the two-terminal method. In Chap. 11, after a study of the possibility to implement an automatic wavelet based fault locator system, it was recommended that the time domain signals must in all cases be examined by a trained person to verify the calculations performed by the automatic fault locator system.

As the data recorded at both cable ends is needed for fault location using any online method, the task of the fault locator unit installed locally at the substation becomes quite simple:

'In the case of a fault, record the three input signals and send the data for central processing'.

This task must be carried out online, and no fault transients can be missed. A sketch of the necessary components if voltage signals are used as input for the fault location unit (FLU) is presented in Fig. 12.1 where a similar setup is used if core or sheath currents are chosen where only the input signal is different.

The acquisition of data is the only task which has to be performed online. After the data is gathered, it can be processed over the course of a few seconds to several minutes depending on the method chosen. The signals gathered at each cable end need to have a common time reference. Such a time reference can be obtained using a GPS-unit. With the GPS protocol, an accuracy of 100 ns is achievable.

12.1 Selection of Equipment

The data acquisition unit including GPS-time synchronisation can be realised using several products already on the market. One of the biggest developers of such is National Instruments (NI). The company is specialised in developing solutions for testing and data acquisition using embedded systems, and their PXI-Express systems

C. F. Jensen, *Online Location of Faults on AC Cables in Underground Transmission Systems*, Springer Theses, DOI: 10.1007/978-3-319-05398-1_12,

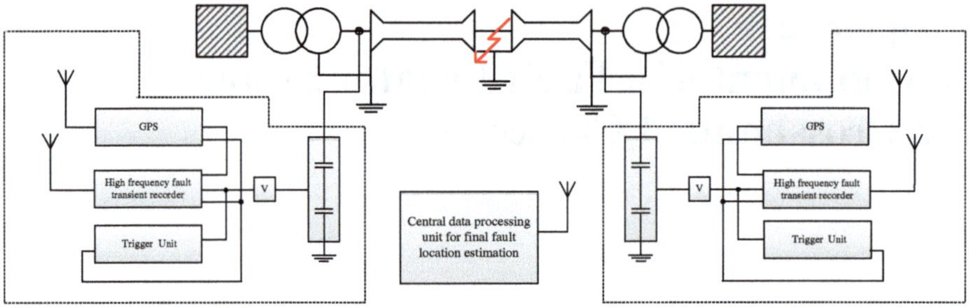

Fig. 12.1 Components included in the proposed fault locator system

Table 12.1 Requirements for a fault locator unit

Requirement	Value
Sampling frequency	≥4 MHz
Time synchronisation error	≤100 ns
Communication delay	≤1 min
Vertical resolution of DAQ	≥10 bit (1024 points)
Buffer size	≥0.02 sec @ 4 MHz

are widely used in industry around the world [1]. For the development process, NI units are chosen.

The NI PXIe system consists of a controller and a number for arbitrarily chosen modules that performers various tasks depending on what solution is necessary. For a fault locator unit for crossbonded cables, the three functions mentioned are data acquisition, time synchronisation and data communication. For development purposes, the following requirements are set up.

With the sampling frequency chosen, the theoretically worst case fault location error is 22 m. For a later implementation, a 10 MHz data acquisition unit can be chosen (9 m theoretically worst case error). All requirements can be met by using the units shown in Table 12.2.

In Fig. 12.2, a single NI-unit with the GPS antenna is shown.

Using NI equipment, the controller can be based on a Windows platform or implemented as a dedicated real time unit. The latter is most optimal for a final implementation, but for development purposes, the Windows based system is simpler to program and use. The communication protocols existing in Windows can be used directly or a self-programmed one can be implemented. The focus of this work is not to develop a secure communication link between the local units and the central data processing location. Therefore, only the development of software needed for the fault location part is described.

Two NI-units are obtained and a Labview program developed for both. Software, which handles the post-fault data treatment is not developed as a part of this thesis as Energinet.dk uses a dedicated system too where there is no access. In the coming section, the software and the verification of the fault locator units is described.

Table 12.2 National instrument units used to realise a fault locator unit for crossbonded cables

Function	Unit
Controller	PXIe-8115 Core i5-2510E 2.5 GHz
Synchronisation	NI PXI-6682H IRIG-B, IEEE 1588 Sync and Time Module
Data acquisition unit	PXIe-6124, 16-bit, 4 MS/s Simultaneous Sampling Multifunction DAQ

Fig. 12.2 National Instrument equipment used to build fault locator unit

12.2 Software Development

As mentioned, the local unit installed at the substation has one main task—to acquire the chosen signals and in the case of a fault, send the data for central data processing. A trigger mechanism must therefore be developed along with the data acquisition routine itself. A software implemented trigger is preferable as no additional hardware units beside the NI equipment are necessary. The software must, however, be developed so it ensures that no data is delayed or missed. To ensure this, a producer/consumer program structure is selected. The structure consists of two main loops running constant and independent of each other. The producer loop's only task is to sample the data contentiously whereafter it is sent to and processed by the consumer loop. As no data is processed in the consumer loop and the queue structural data handling method is selected, no data will be lost due to delays in procession of the data. The only requirement is that on average, the consumer loop is faster than the producer loop's execution time. The program structure is shown in Fig. 12.3.

12.2.1 Producer Loop

The main task of the producer loop is to acquire data from three input channels (three-phases) and time stamp the data with a GPS-based high precision time stamp.

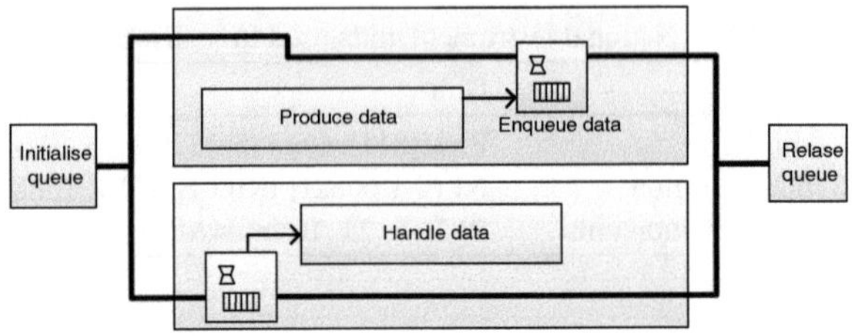

Fig. 12.3 Producer/consumer program structure

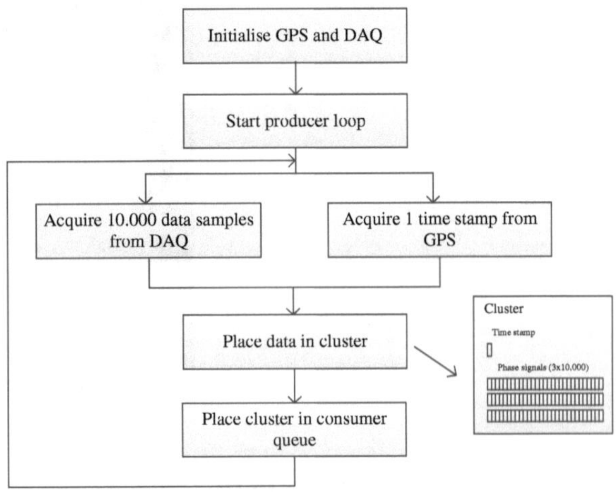

Fig. 12.4 Producer loop program structure

For data sampled at very high frequencies, NI recommends that a certain number of samples are acquired per producer loop iteration. It is found that 10.000 samples per phase ($3 \times 10\,\text{kS}$) per consumer loop iteration is a suitable number at a sampling frequency of 4 MHz. The producer loop iteration time is as a result 2.5 ms.

The GPS-unit cannot generate time stamps with this rate wherefore all samples cannot get a unique time stamp when they are sampled. The time based on the GPS-unit is, however, only synchronised every second while in between, a high precision clock controls the generation of time stamps. The GPS-unit is set to generate one time stamp per producer loop iteration and the additional time stamps for the remaining samples can be extrapolated by adding one sample period per sample. The data is saved in a cluster (NI terminology for a structure) containing one time stamp and three vectors with the sampled data for all three channels. The data is placed in the queue and sent to the consumer loop. The producer loop continues this way infinitely. In Fig. 12.4, the producer program structure is shown.

12.2.2 Consumer Loop

The tasks of the consumer loop is to analyse the data acquired in the producer loop and based on the data, determine if a fault has occurred. In the case of a fault, the

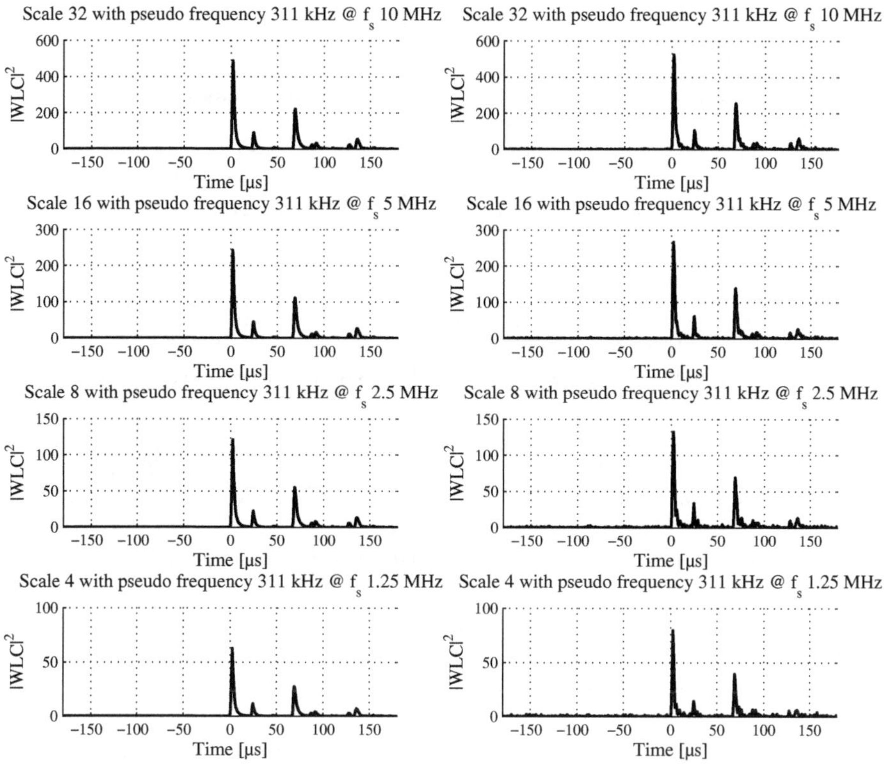

Fig. 12.5 CWCs evaluated at different scales with different sampling frequency keeping a constant pseudo-frequency

sampled data needs to be saved and sent for further data processing. Before the data is sent, communication with the fault locator unit at the other cable end and the relay system ensures, however, that the locally recorded transient signal does in fact contain a fault incident. All events are saved locally on the hard drive for possible system optimisation.

Trigger System

One of the most important tasks of the fault locator unit is to trigger correctly. A trigger must be generated from the FLU itself as the protection system is too slow, resulting in vast data amounts for high frequency sampling. The Wavelet Transform's ability to extract frequency components different than the power frequency signal localised in time can be used to develop a trigger mechanism for a fault locator trigger system. The task is not to identify the exact arrival instance of the wave, but merely to detect its presence. The high scales containing the low frequency component of the analysed signal seems preferable for trigger development for longer cables as low

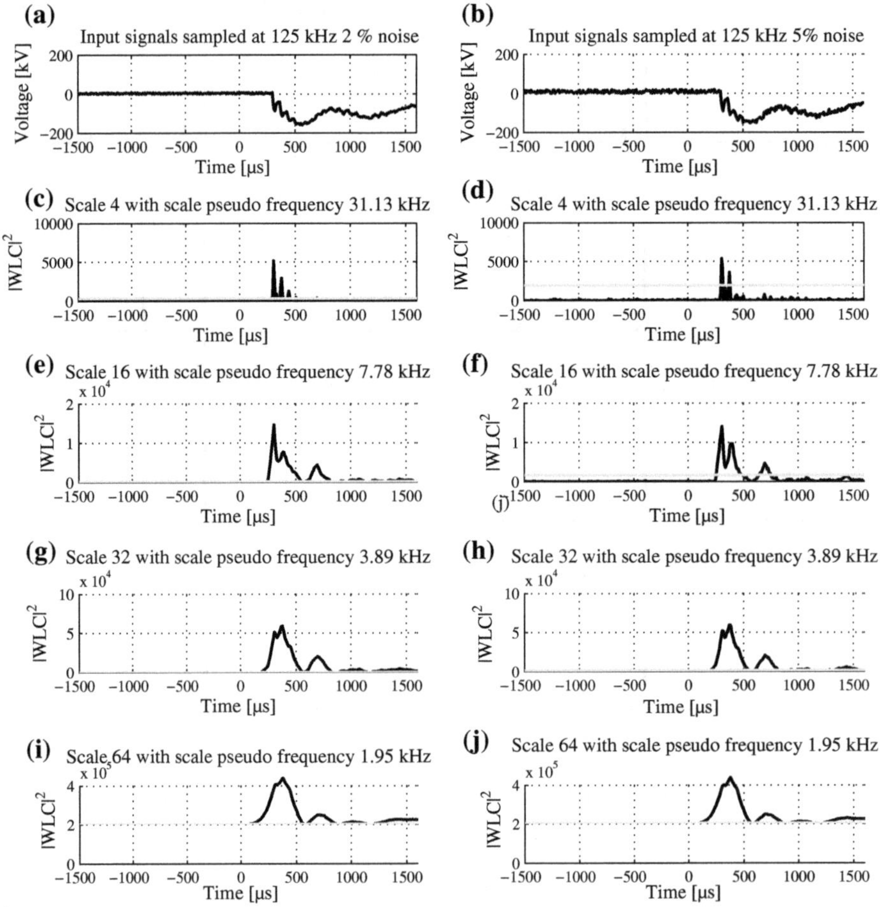

Fig. 12.6 Trigger signal generation at scale 4, 16, 32 and 64 for the core voltage at Terminal B on a 60 km cable with **a** 2 % white noise added and **b** 5 % white noise added

frequency content is expected; for shorter cables, the lower scales can be used with benefit. In order not to have software special developed for each cable system, the combined use of different scales is, thus, a smart solution as the same implementation can work for most cable types.

For a trigger system to work, the WLCs needs to be calculated online at selected scales at a speed higher than one producer loop iteration. This sets a demand for the maximum iteration time for the consumer loop. Instead of using a signal sampled at high frequency and a resulting high scale, a lower scale with lower sampling frequency will result in the same pseudo-frequency and therefore the same frequency band for the two cases. This reduces both the lengths of the analysed signal and the wavelet and thereby the computation effort per iteration. However, information is lost when the signal is downsampled and the arrival of a fault wave is in fact harder to identify. Figure 12.5 shows the Haar WLCs evaluated for the Terminal B signal described in Sect. 12.2.2, at different scales with different sampling frequencies. The scales and sampling frequencies are selected so all pseudo-frequencies are equal ($f_s/a = 311$ kHz where f_s is the sampling frequency and a is the integer scale number). The left column figures are based on clean simulation data where the right column figures are calculated based on an input signal added random white noise with a magnitude of 1.5 % of the non-faulted signal's peak value.

Figure 12.5 shows that as the sampling frequency lowers, the wavelet modulus maxima for the same signal and the same pseudo-frequency lowers as well. The task is, therefore, to select a downsampling rate as low as possible, but still keep it as high so the consumer loop can finish within the time of one producer loop iteration.

The main task of the trigger system is to identify if any type of fault has occurred. To develop such a system, a worst crossbonded cable system on which a trigger must be generated is used. Due to the low frequency content, the most difficult case to generate a trigger for, is a long cable system with one additional cable connected to the same substation as the monitored cable. The 60 km 165 kV cable system with one additional cable connected at both FLT described in Chap. 11 is considered as a worst case system. At Terminal A, the frequency content is high due to the location of the fault close to the terminal. The pre-conditioning of the signal proposed in Chap. 11, is preferable for trigger generation as well. The elimination of the power frequency signal gives lower threshold values and thus better triggering conditions. Furthermore, the WLC needs only to be calculated on one signal instead of three.

A mother wavelet, a downsampling rate and one or more scales must be chosen. All mother wavelets mentioned in the previous section are examined.

From a detailed study, it is found that the Haar wavelet is performing the best for trigger generation. A downsampling rate (n) equal 32 ($f_s/n = 125.000$ kHz) is found to work well. Using scale 4, 16, 32 and 64 with pseudo-frequencies of 31.13, 7.78, 3.89 and 1.95 kHz respectively makes triggering on all examined signals possible. In Fig. 12.6, the trigger generation for the Terminal B voltage on the 60 km cable system described in this section is seen as an example. The left column WLCs are calculated based on a voltage signal added 2 % white noise per phase where 5 % is added to the signal used for the right column WLCs. A worst case of common mode noise is assumed (same noise signal added all three phases).

Figure 12.6 shows that a trigger is generated at all scales for both noise levels. This shows that even on long cables, the incoming fault wave will trigger the fault locator system.

As an additional example, the left collum in Fig. 12.7 shows the trigger generation for a 10 km cable with 5 % noise and the Anholt Case Study I signal at Joint 0 with 2 % additional noise.

Figure 12.7 shows that triggers are generated at all scale for both cases. The four cases presented in this section show that triggering is possible on all realistic cables installed in the future Danish transmission grid if the proposed method is used.

End effects Because the CWT deals with finite-length signals (windowing of continuous signals) errors and bias values often occur at the signal's ends and the

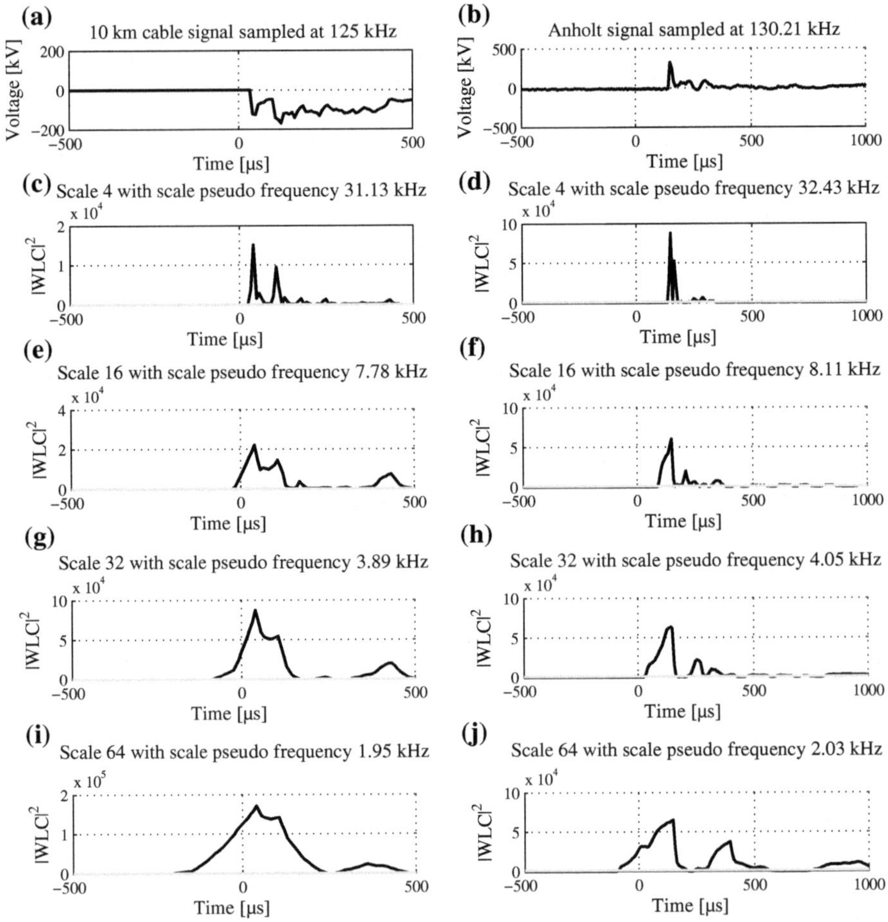

Fig. 12.7 Trigger signal generation at scale 4, 16, 32, and 64 for **a** the core voltage at Terminal B on a 10 km cable signal sampled at 125 kHz and **b** Anholt core voltage at Joint 33 for Case Study I sampled at 130.21 kHz

result is possibly a faulted trigger. The end effect of an arbitrarily chosen signal without pre-conditioning is seen in Fig. 12.8.

Figure 12.8 shows that the threshold comparison will cause a faulted trigger at the signal start and end. A solution to this problem is simply to remove a number of the WLCs so that the end effect is not seen in the remaining. However, if the fault wave arrives at the sample taken between two loops (removed part) an important part of the signal is missing. To deal with this problem, the analysed signal is padded with a part of the signals recorded at the last consumer loop iteration. The part added just needs just to be longer than the part removed for the method to work.

The part of the WLCs affected by the end effect is scale dependent. The WLCs are calculated as the convolution of a flipped-in-time wavelet with the signal. This

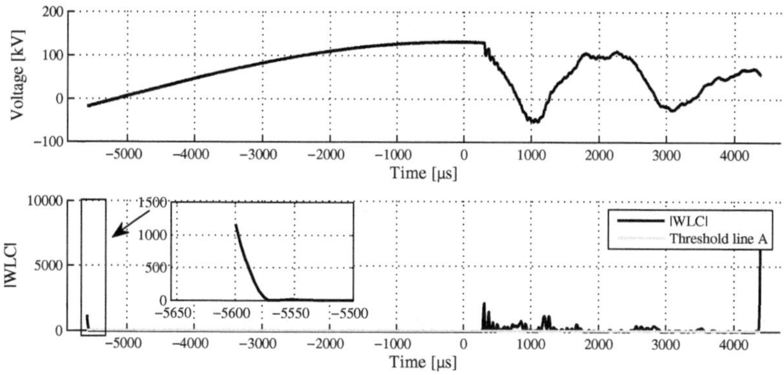

Fig. 12.8 End effect of the Wavelet Transform when taken on a windowed part of a continuous signal

is a common filter operation and the resulting WLCs can be seen as the filter output. The number of elements affected by the end effect at scale a can be shown using convolution theory to be $a - 1$. The length of the filtered signal is $S + a - 1$ where S is the length of the signal. At scale 2, the length of the wavelet is 2, at scale 4, it is 4 and so on. It is normal custom to reduce the length of the filtered signal so it has the same length as the input signal S. This reduces the length of the filtered signal at scale a by a total of $a - 1$ elements, but as the end effect appears in $a - 1$ elements at each signal end, a total of $2a - 2$ elements need to be removed. The length of the resulting signal is thus $S + a - 1 - 2(a - 1) = S - a + 1$ and a minimum of $a - 1$ elements need to be padded. Due to the removal of WLCs, there is a chance that the fault wave is first detected in the following loop iteration. Therefore, the three-phase signals are always saved in a shift register before they are discarded from memory. In case of a fault, the signal at the present and the last iteration are combined in one vector and sent to the central data processing location.

The end effect occurs because the convolution of the wavelet to the analysed signal is carried out at the signal's ends using zero-padding. If the signal is pre-conditioned using the method proposed in this thesis, the non-faulted analysed signal is zero assuming no noise and as a result, the end effects do not occur. However, if noise is present in the signal, end effects of smaller magnitude can occur. This increases the chance of creating wrong triggers and the end points are removed as proposed above.

LabView continuous wavelet transform implementation The Continuous Wavelet Transform is not available in the LabView standard signal processing toolbox and must therefore be implement manually. The CWT can be implemented numerically as shown in Eq. 11.2. Adding the handling of the end effect to the same routine, the CWT at scale 4 is an example implemented as shown in Fig. 12.9.

The input signal is read into the CWT-subroutine and convolved with a flipped-in-time Haar wavelet created at scale a. The convolved output is divided by $\sqrt{a}$ and $a - 1$ points are removed at both signal ends. The signal is squared and the resulting signal is compared to a threshold value point by point. If the square of the i^{it} WLC is

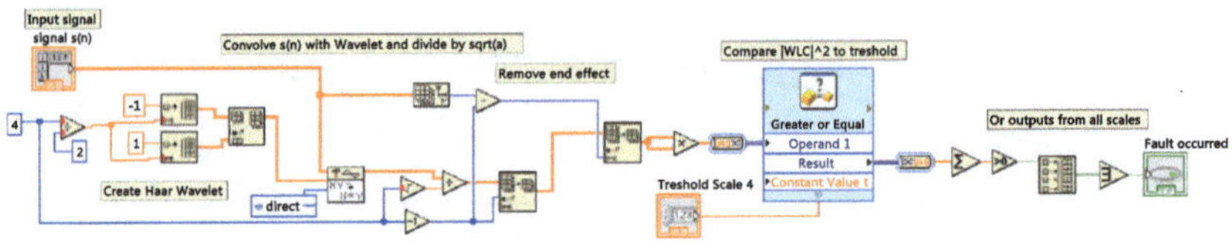

Fig. 12.9 Labview implementation of The Continuous Wavelet Transform and a method to handle the end effect

larger than the threshold value, the Boolean variable 'Fault occurred' is set high. The 'Fault occurred' variable is 'ored' with outputs from the comparison of the CWTs evaluated at other scales to its threshold value.

Threshold value determination Two ways of determining the threshold values are implemented. First, a simple method where the user sets the threshold for each scale is implemented. For a 50 Hz non-faulted noiseless steady state signals, the output from the convolution is a sinusoidal wave and the square of this sine wave is a signal with the double frequency. The WLCs will have their largest value at zero-crossing of the analysed signal. By performing the convolution on a steady state sinusoidal signal, the expected magnitude of the sinusoidal shaped WLCs will be found. After the steady state peak voltage output from the instrument transformers is determined, the thresholds can be calculated and multiplied with a safety margin value to avoid faulted triggers due to noise.

The second method is an online adaptive method that finds the maximum WLC at a given scale for the values calculated in the last consumer loop iteration multiplied by a safety factor. A more advanced algorithm to determine the threshold online can be developed after the first fault locator is installed and the noise at the location can be evaluated.

Trigger backup system In case the Wavelet Based trigger system fails, a backup trigger system is needed. In case of a fault, the differential relay system trips the faulted cable line. A digital signal indicates the occurrence of fault and this signal is monitored by the fault locator by one of the DAQs 24 digital I/O lines. If the relay is tripped, but the Wavelet-based trigger is not activated, the data sampled in the last 20 producer loop iterations (50 ms) is send to central processing. Larger data packages are send, but the data is not lost.

12.2.3 System Verification

The fault locator is tested using a function-based test strategy. The functions of the unit is divided into three main functions. These are:

1. Data acquisition
2. GPS time synchronisation
3. Wavelet Transform based trigger
4. Timing

Data acquisition is tested by sampling a variety of signals on all three input channels and observing the signals represented in the producer loop. No problems are identified.

GPS-time synchronisation is tested by applying 50 Hz 10 V signals on all input channels of both fault locator units. At an arbitrarily instance, a discontinuity is created and the instance of occupance is timed by both fault locators. In all cases, the discontinuity is determined within the GPS maximum time error ($\pm$100 ns).

Wavelet transform based trigger is tested by comparing the output of the self-created CWT-subroutine using known input signals to the output from the MATLAB implemented transform. The outputs are the same at all scales for all cases. As the trigger mechanism is extensively tested using MATLAB and the outputs from the two implementations are equal, the trigger mechanism is correctly verified.

Timing The mean value of the execution time of the consumer loop must be smaller than the value for the producer loop. These times can be measured using the 'Performance and Memory Profiling' tool available in LabView. The producer loop time is 2.5 ms, wherefore the requirement is that the consumer loop execution time is less. Using the profiling tool, the mean value of the consumer loop is measured as 0.08 ms with a standard deviation of 0.02 ms.

12.2.4 Fault Location on Hybrid Lines

The same basic principle for data acquisition as used on crossbonded cables must be used to monitor hybrid lines. Therefore, the same equipment and method as developed in this chapter can be used directly. However, as recommended in Sect. 9.1.1, both core voltage and either core or sheath currents should be monitored. For this reason, six input channels are needed and two DAQ-cards of the proposed type are required in each fault locator unit. There is only an economical issues and has no effect no the functionality of the fault locator system.

12.2.5 Summary

In this chapter, a fault locator system based on travelling waves was designed for crossbonded cables. Two GPS-time synchronised fault locator units were developed based on a National Instrument PXIe platform with Windows 7 as operating system. The fault locator unit sampled three signals from either voltage or currents transformers at a rate of 4 MS/sec. The signals are added point by point to enhance the fault waves and eliminate the power frequency signal. A trigger mechanism was developed based on an online version of the Continuous Wavelet transform. In the event of a fault, the data in the fault locator unit's buffer is saved and if the protection relay for the line monitored is tripped, the data is sent for central processing where the fault location information is extracted. The units are tested using a module-based testing approach and are ready for installation in a Danish substation.

Reference

1. National Instruments. Introduction to pci express. Internet, Vist: 04/05 (2013), http://sine.ni.com/np/app/main/p/ap/global/lang/da/pg/1/sn/n24:PCIE/fmid/2957/

Part IV
Conclusions

Chapter 13
Conclusion

The aim of this work was to develop a fault locator system capable of locating faults with high accuracy on crossbonded cable systems and on hybrid lines. A two-terminal synchronised travelling wave method has been proposed and verified using field measurements conducted on the electrical connection to the offshore wind farm Anholt. The arrival instances of the fault waves are identified based on a combined visual inspection- and Wavelet based method. The proposed fault locator has been constructed based on the principles proposed and will be installed in Danish substations in the future for monitoring of the coming cable-based transmission grid. With the construction of the fault locator system, all problems formulated in the problem formulation are answered fully.

13.1 Summary of the Thesis

Beside the preliminaries and this conclusion chapter, the thesis has been divided into two main parts; one dealing with fault location using impedance-based methods and one dealing with fault location using travelling wave methods. In each part, the problems identified in the problem formulation are treated:

1. To study the steady state and transient behaviour of a crossbonded cable system under fault conditions.
2. To analyse the correct modelling and simulation techniques of fault location studies on crossbonded cable systems.
3. To identify and possibly improve the best suited fault location method for crossbonded cables and hybrid lines.
4. To study the necessary measuring equipment required for sufficient accurate fault location in crossbonded cable networks and on hybrid lines.
5. To verify the proposed fault location method using field measurements conducted on a real-life crossbonded cable.
6. To develop a prototype of the proposed fault locator unit.

C. F. Jensen, *Online Location of Faults on AC Cables in Underground Transmission Systems*, Springer Theses, DOI: 10.1007/978-3-319-05398-1_13,

13.1.1 Summary of the Impedance-Based Fault Location Methods for Crossbonded Cables

Based on a literature study, it was found that very little research have been conducted on fault location on crossbonded cables using impedance-based methods. The series impedance matrix which together with other zero-sequence parameters of the system dominates the behaviour of the fault loop impedance was presented. It was found that an analytical expression for the exact fault loop impedance is very difficult to derive and several assumptions has to be made. For fault location problems where the accuracy is a major concern, this is a problem.

A highly detailed model of a two-major section crossbonded cable was implemented in DIgSILENT Power Factory. It was found that the fault loop impedance is not linear dependent on the fault location due to the transposing of the sheath at the crossbondings. This is a unique problem for crossbonded cables and a major difficulty for impedance-based fault location methods, as a linear- or a relationship governed by hyperbolic functions is expected to relate the fault loop impedance to the fault location.

Because the return currents shares common paths back to the sources on crossbonded cables, the fault loop impedance in case of bolted fault with double-sided infeed is affected when measured from a single side. This is also a unique problem for crossbonded cables, and, it is further problematic, if no compensation method is applied.

From a study on the influencing parameters, it was found that the field grounding resistances at the ends of major sections and substation grounding resistances do not affect the fault loop impedance significantly. This is because almost the entire fault current returns to the source in the sheath system due to the high impedance path provided by the ground.

In case of a fault resistance different from zero between core and sheath, a strong effect in both the real and imaginary parts of the fault loop impedance for double-sided infeed is seen. As the fault loop impedance per unit length is small on cable systems, this is a problem as only a small part of the measured impedance is directly related to the distance to fault.

13.1.2 Summary of Fault Location on Hybrid Lines Using Impedance-Based Methods

Impedance-based fault location on hybrid lines is very difficult to implement for practical cases. The fault loop impedance is almost four to five times as large on OHL systems compared to cable systems in touching trefoil and flat formation. Small errors in the OHL's parameters will therefore give relatively large errors when calculating the fault location in the cable section and the method therefore becomes applicable in theory only.

13.1.3 Summary on Fault Location Using Neural Networks

It was examined whether neural networks could be trained to perform fault location on crossbonded cable systems using data from state-of-the-art simulation programs (DIgSILENT in this case). A vast set of impedance-based field measurements were conducted on the electrical connection to the offshore wind farm Anholt. All possible types of short circuits were applied on several combinations of the system and it was concluded that the simulation program predicts the behaviour of the crossbonded cable system well, but the precision was low. Deviations up till 200 % on the real part and 61 % of the imaginary part of the fault loop impedance were found. The training data for the neural network will therefore not be valid for real-life behaviour, wherefore the method is not practically applicable. The main conclusion of part I is therefore:

> *An impedance-based method is not well suited for fault location on neither pure crossbonded cables nor on hybrid lines.*

13.1.4 Summary of the Travelling Wave-Based Fault Location Methods for Crossbonded Cables

A literature study was carried out to examine the use of a travelling wave-based fault locator for crossbonded cables. It was found that very little research had been carried out and the few references found are only focused on very short cables. Furthermore, fault wave propagation on crossbonded cables is not a field studied in depth. It was found that more research was needed before any conclusions could be made.

Wave propagation on three-phase, single-core cable systems was examined first. The modal domain was introduced and the modal decomposition of cable systems in touching trefoil and flat formation were determined. It was found that coaxial waves are the fastest waves existing on the system at any frequency of interest and that only coaxial waves are created when an ideal fault is applied between core and sheath anywhere on the cable system.

Wave propagation on a crossbonded cable was examined and it was found that intersheath and ground mode waves are created at the crossbondings every time a coaxial wave arrives. For fault location purposes, this means that the first intersheath mode wave to arrive at the fault locator terminal has not travelled from the fault location, but was created at the crossbonding closest to the fault locator terminal.

From a study of fault wave propagation on longer crossbonded cable systems, it was found that a part of the fault wave travelling between core and sheath on one cable is reflected onto the two other cables due to the interconnection between the sheaths at the crossbondings. After 4–5 major sections, fault waves of equal magnitude exist on the three cables. The reduction in magnitude of the initial fault wave is a problem unique for crossbonded cables and can make fault location more difficult in a noisy environment.

The Use of the Single- and Two-Terminal Fault Location Method on Crossbonded Cables

Actual real-life fault location on a three major section crossbonded cable was examined. It was found that the two-terminal method gives very accurate results when the coaxial modal velocity is used to define the wave velocity and extracting the arrival instance of the fault wave directly in the time domain. Core voltages, core current or sheath current signals can all be used as input for the fault locator, whereas the choice of signal type is dependent on the number of additional cables or OHLs connected at the substation where the monitored cable is connected. As more lines are connected, the current signal becomes more advantageous to use as input, whereas, if a single line/single transformer configuration is used, only core voltage signals are applicable.

The single-terminal method is very problematic to use on crossbonded cables as all modal wave types are created when a travelling wave meets a crossbonding point. The single-terminal method relies on capturing the second wave from the fault location, but due to the vast number of waves travelling in the system, the second wave from the fault location becomes very hard to distinguish from the waves created at the crossbondings.

A sheath current signal method was examined, but it was concluded that this method is not applicable in practice as the location of faults in the minor sections closest to the fault locator terminal could not be determined using the method.

Parameters Influencing a Two-Terminal Fault Location Method for Fault Location on Crossbonded Cables

The influence of the most important parameters influencing the two-terminal method was examined.

The wave velocity on a single-core cable system is frequency dependent. The highest frequency components travel the fastest which makes the initial wave seem spread out and the exact arrival instance becomes more difficult to determine exactly. However, the effect is minor and no correction is necessary on cable systems less than 50 km (98.6 % of all Danish lines).

The effect when the fault wave meets the busbars at different types of substations was examined. It was found that for outdoor substations of a certain size, it becomes advantageous to always use voltage waves as input for a fault locator because the incoming fault wave is almost doubled when the waves meet the busbars. The time it takes the inter-station waves to settle is larger than the sampling time used for travelling wave-based fault location, and the inter-station waves are therefore picked up. At the cable termination at the substation, core current and sheath current waves are reduced to approximately 15 % relative to the incoming fault wave due to a negative current reflection coefficient. For GIS stations, where the busbars are short and the surge impedance is more comparable to the surge impedance of the high voltage coaxial cable, the signal behaviour is determined by the number of additional lines connected to the GIS.

Internal and joint faults on cables with extruded insulation will always occur near peak voltage where the insulation is stressed the most. However, some deviation can be expected and the wave flow in case of faults with fault inception angles of 25°, 30° and 45° are simulated, and the fault wave arrival instance at the fault locator terminals can be identified with no loss in precision.

The sensitivity of the coaxial modal velocity on cables and cable system parameters was examined. It was found that the velocity is dependent on both temperature and frequency, but the effects are minor under normal operation conditions for the cable. The influence of the semi-conducing layers was examined with special attention to their frequency dependent permittivity and their effect on the coaxial attenuation. Based on existing literature, it could be concluded that the effects are minor as well and no compensation is needed. Field measurements were later carried out to confirm these claims.

The use of today's instrument transforms for fault location purposes was examined. Capacitive voltage transformers with access to the neutral of the capacitive voltage divider and Rogowski coils for measuring sheath currents can be used with high precision. Use of inductive voltage and current transformers is recommended by several authors, but the promised bandwidth is lower for both transformer types compared to capacitive voltage transformers and Rogowski coils. Until more experience is gathered, it is therefore recommended always to use a reference signal from a capacitive voltage transformer or a Rogowski coil if inductive transformers are used.

Hybrid Lines

A method developed for fault location on hybrid DC-lines was re-designed for hybrid systems with crossbonded cables and OHLs. The method works for any combination of line segments and was verified using simulations on a hybrid line consisting of a 3 km 1 major section crossbonded cable and a 40 km OHL. Furthermore, the method was tested using a model of the electrical connection to the Danish offshore wind farm Horns Reef 2. It was found that due to damping primarily on the submarine cable section, it is beneficial to install an additional fault locator at the reactor station close to the start of the submarine cable. This reduces the total length of the monitored line and increases the accuracy considerably.

Based on the findings in the first chapters, it was concluded that the two-terminal method was useable if a constant coaxial wave velocity could be used as input for the fault locator and if the degree of damming predicted by current simulation models was representative for real-life coaxial damping. To verify these claims, field measurements on the electrical connection to the Anholt cable were conducted.

Travelling Wave-Based Field Measurements for Verification of Fault Location Methods for Crossbonded Cables

Field measurements were conducted on parts of the electrical connection to the Anholt offshore wind farm. Faults were applied on the 38 km system and the transient response was measured at two terminals at the cable's ends. The measurements were time-syncronised and it was found that a very accurate estimation of the fault location could be obtained using the proposed method. Methods for measuring the coaxial wave velocity were set up and the coaxial wave velocity on the Anholt cable was determined using these methods. It was verified that a constant coaxial wave velocity could be used and the coaxial attenuation predicted by current cable models is in line with results obtained on the Anholt cable. Based on the results of the field measurements, it was concluded that:

> *Fault location using a synchronised two-terminal method is applicable on crossbonded cables with use of the coaxial wave velocity and fault signals analysed directly in the time domain.*

The Wavelet Transform and Fault Location on Crossbonded Cable Systems

The use of the Wavelet Transform for fault location on crossbonded cables was examined. It was found that the transform can determine the fault location on shorter cable systems with high accuracy (approximately less than 20 km). For longer cables, it is proposed to combine the use of the automatic wavelet transform with a visual inspection of the time domain signals. In all cases, this increases the accuracy of the fault location estimation and reduced the chance of acting on a faulted estimation by the Wavelet transform.

Development of a Fault Locator System for Crossbonded Cables

A fault locator system capable of locating faults on crossbonded cables was developed and realised in practice.

Two fault locator units, which have to be installed at each terminal at each cable end, will acquire input signals in case of a fault on the monitored line. The units are permanently installed and constantly monitor the signals. They will trigger and save their inputs in case of a fault. After the data is recorded locally, it is sent to central processing at a remote location where the fault location can be estimated based on the combined use of the Wavelet-based method and the visual inspection method developed.

Core voltage, core currents or sheath currents can freely be chosen as input to the fault locator units. The inputs can be different at each cable end where the choice will depend on the station and the number of additional lines connected to the station. The fault locator units can be used to monitor both pure crossbonded cables and hybrid lines as it is only the post-fault data treatment which is different for the two system configurations.

The fault locator unit is realised using National Instrument equipment. A 4 MS data acquisition card and a GPS-based time-synchronisation card are used. The necessary software was developed in Labview on a Windows 7 platform. A Wavelet based trigger system capable of triggering on all realistic fault signals was developed with the use of a signal pre-condition technique developed especially for crossbonded cables.

The functionality of the units were verified using both simulated and field measurement data and was found to operate as expected.

13.2 Contributions

The main contributions of the thesis are:

- A detailed study on the fault loop impedance and the parameters influencing it on crossbonded cables.
- A study on the use of impedance-based fault location method on hybrid lines.
- Field measurements showing the possible use of state-of-the-art simulation program for providing training data for neural networks.
- A study on fault wave propagation on crossbonded cables.
- A study on the best suited travelling wave-based fault location method for crossbonded cables.
- A study on the parameters affecting the ability of the two-terminal method to locate faults on crossbonded cables.
- Field measurements of fault transients on an installed crossbonded cable and the analysis of these for fault location purposes.
- The use of the Wavelet transform for fault location on crossbonded cables.
- The development of a fault locator system capable of locating faults on crossbonded cables and on hybrid lines.

13.3 Future work

Most of the future work is related to the practical issues of implementing the fault locator.

13.3.1 Signal Conditioning

Depending on the output from the chosen transducer type, some signal conditioning can be required. The DAQ takes a ± 10 V signal as input wherefore some conversion can be necessary. Signal conditioning block are produced by National Instruments and can be easily mounted on the DAQ-unit.

13.3.2 Practical Installation

The cable system on the secondary side of the transducer to the location of the fault locator will influence the accuracy. Methods for determining the delay need to be developed and a compensation routine implemented in the fault location.

13.3.3 Instrument Transformer

More studies are needed on the inductive instrument transformers before they can be used without a reference signal from a capacitive voltage transformer or a Rogowski current coil. A fault locator unit can be installed in a substation where switching operations are expected. Such operations can result in high frequency components in the signals and these can then be measured with several transducer types if these are installed at the same substation. A comparison between the transducers ability to reflect the high frequency components in their secondary circuit can then be made and recommendations can be put up.

13.3.4 Wavelet-Based Trigger Mechanism

The method and settings for the trigger mechanism needs to be studied in more detail. The noise level in substations can vary and the effect of the triggering system must be examined. This can only be done after the fault locator unit is installed in a substation. Furthermore, a strategy for preventing faulted triggers should be implemented. Integration with the existing line protection system should be carried out.

Appendix A
Impedance-Based Fault Location Measurement Results

A.1 Results

A.1.1 Case study 2

Table A.1 Measured and calculated fault impedances for the cable sections from Joint 33 to Joint 27 with the fault location at Joint 27 (Case Study 2)

Fault type	Impedance [Ω]	Measured [Ω]		Calculated [Ω]	
Phase ABC/S	Z_{1A}	1.57 ∠77.58	0.34+j1.53	1.42 ∠74.40	0.38+j1.37
(positive-seq)	Z_{1B}	1.30 ∠84.09	0.13+j1.29	1.20 ∠84.00	0.13+j1.19
	Z_{1C}	1.54 ∠91.20	−0.03+j1.54	1.37 ∠93.60	−0.09+j1.37
	Z_1	1.47 ∠84.29	0.15+j1.45	1.33 ∠84.00	0.14+j1.31
Phase AB/S	$Z_{f,A}$	1.11 ∠76.05	0.27+j1.08	1.42 ∠74.40	0.38+j1.37
	$Z_{f,B}$	1.88 ∠68.90	0.68+j1.75	1.20 ∠84.10	0.12+j1.19
Phase BC/S	$Z_{f,B}$	1.00 ∠72.18	0.31+j0.95	1.20 ∠84.00	0.13+j1.19
	$Z_{f,C}$	2.25 ∠75.64	0.56+j2.18	1.37 ∠93.56	−0.09+j1.37
Phase CA/S	$Z_{f,A}$	2.60 ∠58.20	1.37+j2.21	1.42 ∠74.40	0.38+j1.37
	$Z_{f,C}$	1.13 ∠76.23	0.27+j1.10	1.37∠93.56	−0.09+j1.37
Phase A/S	$Z_{f,A}$	1.30 ∠67.54	0.50+j1.20	1.16 ∠74.90	0.30+j1.12
Phase B/S	$Z_{f,B}$	1.12 ∠64.09	0.49+j1.01	1.01 ∠71.60	0.32+j0.96
Phase C/S	$Z_{f,C}$	1.30 ∠67.36	0.50+j1.20	1.15 ∠72.40	0.35+j1.10
	Z_{0A}	1.30 ∠21.09	1.22+j0.47	0.92 ∠56.20	0.51 0.76
	Z_{0B}	1.30 ∠21.57	1.21+j0.48	0.93 ∠55.70	0.52 0.77
Zero-sequence	Z_{0C}	1.29 ∠21.87	1.20+j0.48	0.91 ∠55.82	0.51 0.75
	Z_0	1.30 ∠21.51	1.21+j0.48	0.92 ∠55.91	0.52 0.76

C. F. Jensen, *Online Location of Faults on AC Cables in Underground Transmission Systems*, Springer Theses, DOI: 10.1007/978-3-319-05398-1,

A.1.2 Case Study 3

Table A.2 Measured and calculated fault impedances for the cable sections from Joint 0 to Joint 33 with the fault location at Joint 27 (Case Study 3)

Fault type	Impedance [Ω]	Measured [Ω]		Calculated [Ω]	
Phase ABC/S	Z_{1A}	7.16 ∠77.99	1.49+j7.01	6.49 ∠76.38	1.53+j6.31
(positive-seq)	Z_{1B}	5.94 ∠85.12	0.50+j5.91	5.35 ∠82.70	0.68+j5.31
	Z_{1C}	6.99 ∠91.89	−0.23+j6.99	6.36 ∠91.50	−0.17+j6.36
	Z_1	6.70 ∠85.00	0.59+j6.64	6.07 ∠83.53	0.68+j6.03
Phase AB/S	$Z_{f,A}$	5.11 ∠78.76	1.00+j5.01	4.74 ∠74.68	1.25+j4.57
	$Z_{f,B}$	8.17 ∠69.02	2.92+j7.62	7.92 ∠70.50	2.64+j7.47
Phase BC/S	$Z_{f,B}$	4.59 ∠74.22	1.50+j3.94	4.22 ∠69.08	1.51+j3.94
	$Z_{f,C}$	9.87 ∠73.92	1.78+j9.43	9.46 ∠70.64	3.13+j8.93
Phase CA/S	$Z_{f,A}$	11.18 ∠56.50	6.17+j9.32	11.26 ∠60.90	5.51+j9.89
	$Z_{f,C}$	5.11 ∠78.93	0.98+j5.01	5.09 ∠74.75	1.29+j4.72
Phase A/S	$Z_{f,A}$	6.00 ∠70.37	2.01+j5.65	5.67 ∠66.69	2.24+j5.21
Phase B/S	$Z_{f,B}$	5.15 ∠66.36	2.07+j4.72	4.80 ∠61.07	2.32+j4.20
Phase C/S	$Z_{f,C}$	5.88 ∠69.70	2.04+j5.51	5.61 ∠65.78	2.30+j5.12

A.1.3 Case Study 4

Table A.3 Measured and calculated fault impedances for the cable sections from Joint 33 to Joint 0 with the fault location at Joint 27 (Case Study 4)

Fault type	Impedance [Ω]	Measured [Ω]		Calculated [Ω]	
Phase ABC/S	Z_{1A}	1.56 ∠77.53	0.34+j1.52	1.42 ∠74.70	0.37+j1.37
(positive-seq)	Z_{1B}	1.28 ∠84.15	0.13+j1.28	1.20 ∠84.00	0.13+j1.19
	Z_{1C}	1.52 ∠91.03	−0.03+j1.52	1.37 ∠93.60	−0.09+j1.37
	Z_1	1.46 ∠84.23	0.15+j1.44	1.33 ∠84.10	0.14+j1.31
Phase AB/S	$Z_{f,A}$	1.11 ∠76.11	0.27+j1.07	1.12 ∠81.90	0.16+j1.11
	$Z_{f,B}$	1.86 ∠68.96	0.67+j1.73	1.46 ∠70.15	0.50+j1.37
Phase BC/S	$Z_{f,B}$	0.99 ∠72.30	0.30+j0.95	1.02∠79.00	0.19+j1.00
	$Z_{f,C}$	2.22 ∠75.46	0.56+j2.15	1.72∠78.80	0.33+j1.69
Phase CA/S	$Z_{f,A}$	2.59 ∠57.86	1.38+j2.20	2.21 ∠56.97	51.20+j1.85
	$Z_{f,C}$	1.13 ∠76.23	0.27+j1.09	1.12∠81.74	10.16+j1.11
Phase A/S	$Z_{f,A}$	1.30 ∠67.67	0.50+j1.21	1.20∠76.70	0.28+j1.17
Phase B/S	$Z_{f,B}$	1.12 ∠64.15	0.49+j1.01	1.05∠72.90	0.31+j1.00
Phase C/S	$Z_{f,C}$	1.30 ∠67.36	0.50+j1.20	1.17∠73.90	0.32+j1.12

A.1.4 Case Study 5

Table A.4 Measured and calculated fault impedances for the cable sections from Joint 0 to Joint 33 with the fault location at Joint 33 (Case Study 5)

Fault type	Impedance [Ω]	Measured [Ω]		Calculated [Ω]	
Phase ABC/S	Z_{1A}	8.75 ∠77.64	1.87+j8.55	7.95∠76.32	1.88+j7.721
(positive-seq)	Z_{1B}	7.27 ∠84.78	0.66+j7.24	6.53∠82.82	0.82+j6.481
	Z_{1C}	8.56 ∠91.66	−0.25+j8.56	7.80 ∠91.72	−0.23+j7.8036
	Z_1	8.19 ∠84.69	0.76+j8.11	7.43∠83.62	0.82+j7.333
Phase AB/S	$Z_{f,A}$	6.27 ∠77.88	1.32+j6.13	5.78 ∠74.70	1.53+j5.587
	$Z_{f,B}$	10.12 ∠75.40	2.55+j9.79	9.68 ∠70.50	3.23+j9.127
Phase BC/S	$Z_{f,B}$	5.66 ∠73.56	1.60+j5.42	5.16∠69.00	1.85+j4.8294
	$Z_{f,C}$	12.13 ∠74.87	3.16+j11.71	11.57∠79.30	2.15+j11.3793
Phase CA/S	$Z_{f,A}$	13.86 ∠57.04	7.54+j11.63	13.92 ∠60.23	56.91+j12.08
	$Z_{f,C}$	6.30 ∠78.17	1.29+j6.16	5.95∠74.60	11.58+j5.74
Phase A/S	$Z_{f,A}$	7.39 ∠69.70	2.56+j6.93	6.91∠66.68	2.74+j6.351
Phase B/S	$Z_{f,B}$	6.38 ∠65.67	2.63+j5.81	5.86∠61.19	2.82+j5.130
Phase C/S	$Z_{f,C}$	7.25 ∠69.08	2.59+j6.77	6.66∠66.00	2.71+j6.082
Zero-sequence	Z_{0A}	6.96 ∠25.84	6.26+j3.03	7.34 ∠22.90	06.76+j2.86
	Z_{0B}	7.00 ∠24.63	6.37+j2.92	7.25 ∠21.30	06.75+j2.63
	Z_{0C}	6.78 ∠26.10	6.09+j2.98	7.24 ∠23.50	06.64+j2.89
	Z_0	6.91 ∠25.53	6.24+j2.98	7.28∠22.57	06.72+j2.79

A.1.5 Case Study 6

Table A.5 Measured and calculated fault impedances for the cable sections from Joint 33 to Joint 0 with the fault location at Joint 0 (Case Study 6)

Fault type	Impedance [Ω]	Measured [Ω]		Calculated [Ω]	
Phase ABC/S	Z_{1A}	8.73∠77.41	1.90+j8.52	7.88 ∠75.50	1.97+j7.63
(positive-seq)	Z_{1B}	7.27∠84.72	0.67+j7.23	6.63 ∠82.80	0.83+j6.58
	Z_{1C}	8.51∠91.95	−0.29+j8.51	7.75 ∠92.50	−0.34+j7.74
	Z_1	8.17∠84.69	0.76+j8.09	7.42 ∠83.60	0.82+j7.32
Phase AB/S	$Z_{f,A}$	6.24∠77.88	1.31+j6.10	5.98 ∠75.50	1.50+j5.79
	$Z_{f,B}$	10.10 ∠69.33	3.56+j9.45	9.39 ∠71.80	2.93+j8.92
Phase BC/S	$Z_{f,B}$	5.64∠73.56	1.60+j5.41	5.41 ∠70.70	1.79+j5.11
	$Z_{f,C}$	12.08 ∠74.81	3.17+j11.66	11.04 ∠80.90	1.75+j10.90
Phase CA/S	$Z_{f,A}$	13.80 ∠57.32	7.45+j11.61	13.37 ∠60.30	56.62+j11.61
	$Z_{f,C}$	6.26∠78.29	1.27+j6.13	6.07 ∠76.40	11.43+j5.90
Phase A/S	$Z_{f,A}$	7.36 ∠69.63	3.96+j9.80	7.02 ∠68.70	3.27+j9.47
Phase B/S	$Z_{f,B}$	6.36 ∠65.80	2.61+j5.80	5.98 ∠63.50	2.67+j5.35
Phase C/S	$Z_{f,C}$	7.24 ∠69.08	2.58+j6.76	6.82 ∠68.25	2.53+j6.33
	Z_{0A}	6.91 ∠25.58	6.23+j2.98	6.93∠28.30	06.10+j3.29
	Z_{0B}	6.95 ∠24.49	6.33+j2.88	6.95 ∠26.90	06.20+j3.14
Zero-sequence	Z_{0C}	6.76 ∠25.84	6.09+j2.95	6.82∠28.70	05.98+j3.28
	Z_0	6.88 ∠25.30	6.22+j2.94	6.90∠27.97	06.09+j3.23

Appendix B
Power System Components Used in the Thesis

B.1 PSCAD Models

In this section the PSCAD/EMTDC implementation of models used in the thesis are described.

B.1.1 Feeder System

A standard feeder system is used in this thesis. The system consists of a 400 MVA 410/167 kV auto transformer and a short circuit impedance. For high frequency studies, the grid behind the transformer is of very little importance and a passive lumped represent of the network is used. The system is shown in Fig. B.1.

The 50 Hz transformer model available in PSCAD/EMTDC is used with a Π of capacitors used to represent the high frequency response of the transformer. This method is described in [1] and the values are taken from [2] (Fig. B.2).

B.1.2 165 kV Case Study Cable

The 165 kV case study ABB cable is modelled using the frequency dependent phase model (Universial line model). The parameters are presented in the thesis in Table 4.2. The PSCAD/EMTDC implementation of the model and the configuration of the cables is shown in Fig. B.3.

The fitting error of the characteristic admittance is 0.195 % and 11 poles are used. Four group delays are defined for the propagation function. These have fitting errors of 0.06793, 0.09666, 0.07645 and 0.03230 %. The fitting error for the phase of the propagation constant is 0.1659 %.

C. F. Jensen, *Online Location of Faults on AC Cables in Underground Transmission Systems*, Springer Theses, DOI: 10.1007/978-3-319-05398-1,

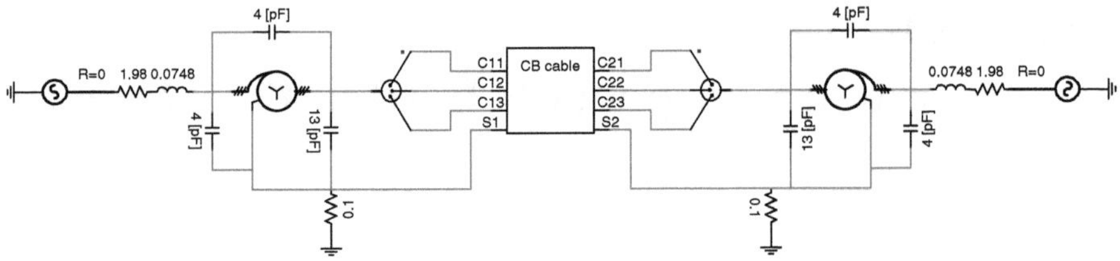

Fig. B.1 Modelling of standard feeder system in PSCAD/EMTDC

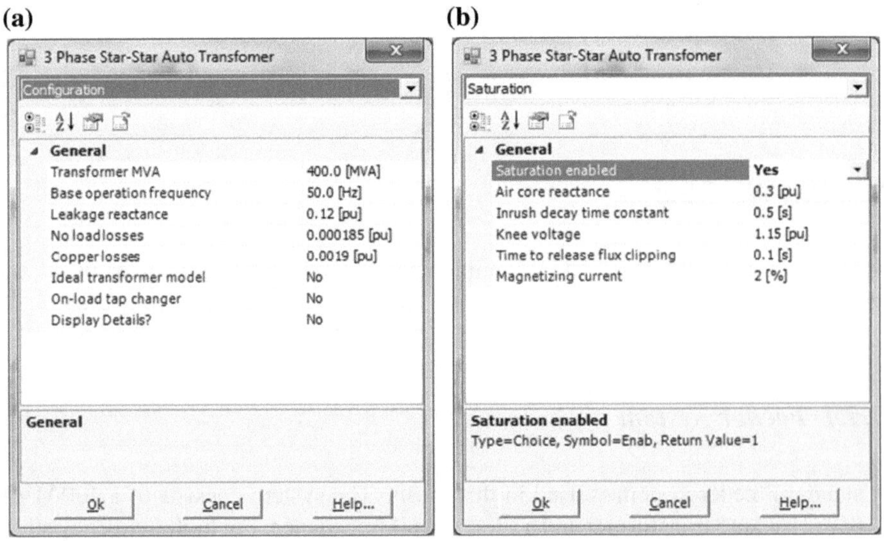

Fig. B.2 **a** 3-phase auto transformer configuration and **b** setup of saturation characteristics of auto transformer

B.1.3 165 kV Case Study Overhead Line

A 165 kV overhead line is used in the study of fault location on hybrid lines in Sects. 4.4 and 9.1. Figure B.4, shows the configurations of OHL model.

Figure B.5a and b shows the configuration of the line structure and the configurations for the conductors.

Figure B.6, shows the configuration for the ground wires.

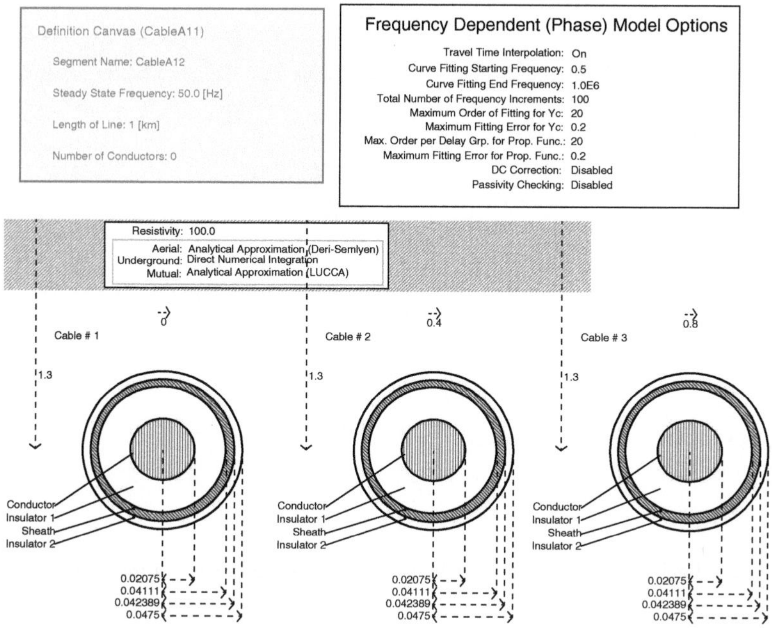

Fig. B.3 Configuration of cable system model

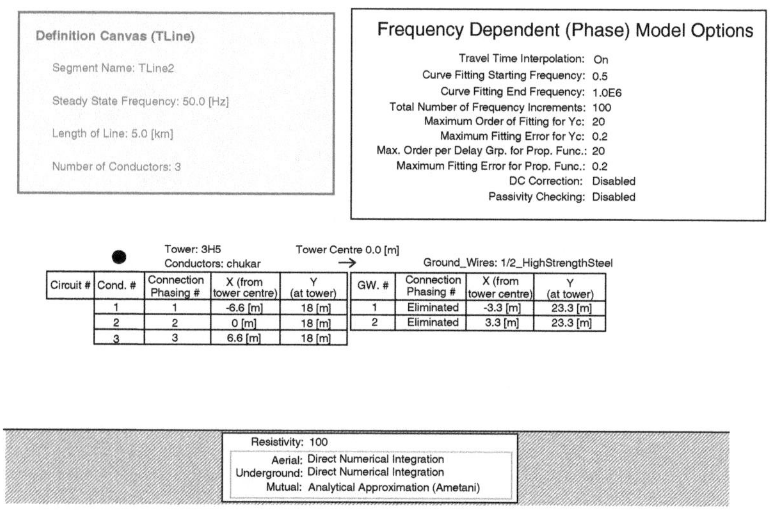

Fig. B.4 Configurations of OHL model

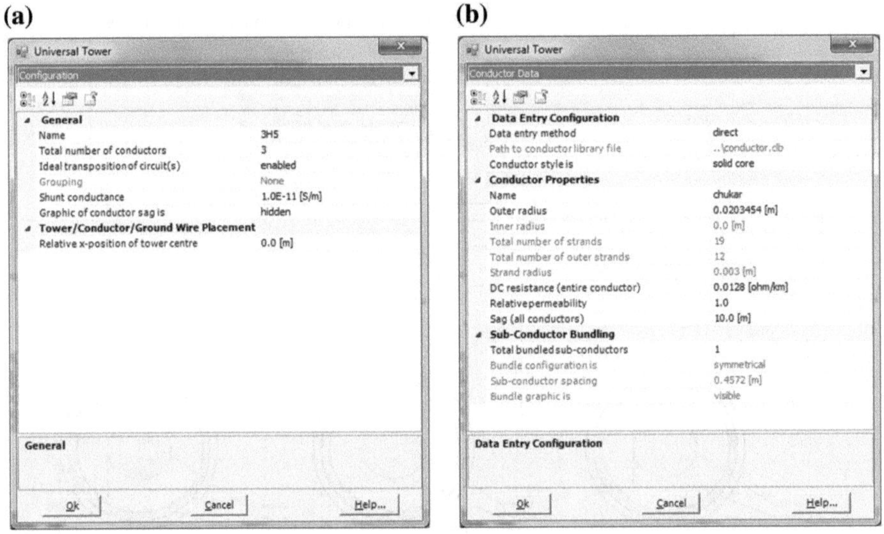

Fig. B.5 Configuration of the **a** line structure and **b** the configurations for the conductors

Fig. B.6 Configuration of ground wires

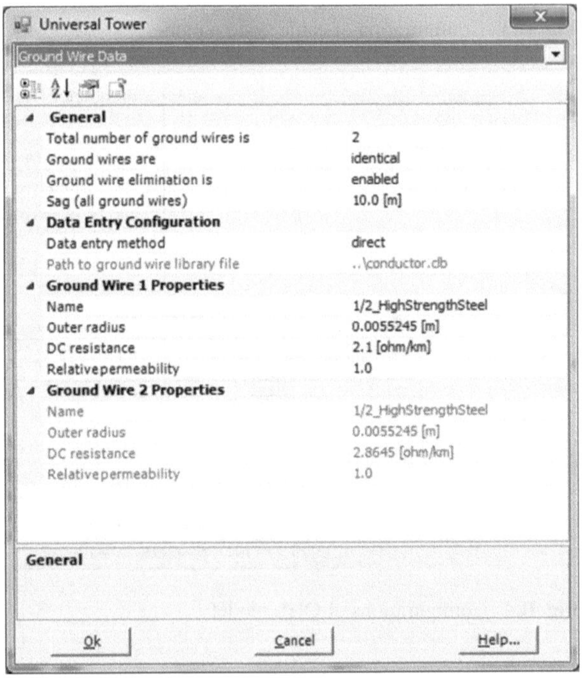

References

1. A. Greenwood, Electrical Transients in Power Systems, 2nd edn. (Wiley, 1991) ISBN 0-471-62058-0. http://www.amazon.com/Electrical-Transients-Power-Systems-Greenwood/dp/0471620580
2. IEEE Guide for the Application of Transient Recovery Voltage for AC High-Voltage Circuit Breakers, IEEE Std C37.011-2011 (Revision of IEEE Std C37.011-2005, Nov 28 2011), pp. 1–97. doi:10.1109/IEEESTD.2011.6093699, http://ieeexplore.ieee.org/xpl/login.jsp?tp=&arnumber=6093699&url=http%3A%2F%2Fieeexplore.ieee.org%2Fstamp%2Fstamp.jsp%3Ftp%3D%26arnumber%3D6093699

Appendix C
Seven-Step Impedance Measuring Method

A seven step measuring procedure developed for determining the positive and zero sequence impedance of OHLs and cables.

C.1 Positive Sequence Impedance

The positive sequence impedance is determined as a mean value of the loop impedance between the three phases as shown in Fig. C.1.

The impedance between phase A and B, between phase A and C and between phase B and C are determined as shown in Fig. C.1. The positive sequence impedance is calculated based on these measurements as where the subscript M denoted a measured impedance and C denotes a calculated impedance and terminal E represents all possible return paths back to the source (sheath, substation grid etc.)

$$Z_{1,C} = \frac{Z_{AB,M} + Z_{AC,M} + Z_{BC,M}}{3 \cdot 2} \qquad [\Omega] \tag{C.1}$$

C.2 Zero Sequence Impedance

The zero sequence system is defined as shown in Fig. C.2.

The line impedances $Z_{A,C}$, $Z_{B,C}$ and $Z_{C,C}$ are assumed equal and the impedance $Z_{E,C}$ represents the return path for the current. In the case of a crossbonded cable system, this is the return path provided by the combined sheath system and the ground. The grounding resistances at the ends of major section and the grounding resistance at the substation(s) are included in the impedance.

The power available will depend on the source. It is proposed to measured the zero sequence impedance in one, a single cable is energised at the time and for method two (Fig. C.3d), all three phases are energised at the same time. In the latter case, the

C. F. Jensen, *Online Location of Faults on AC Cables in Underground Transmission Systems*, Springer Theses, DOI: 10.1007/978-3-319-05398-1,

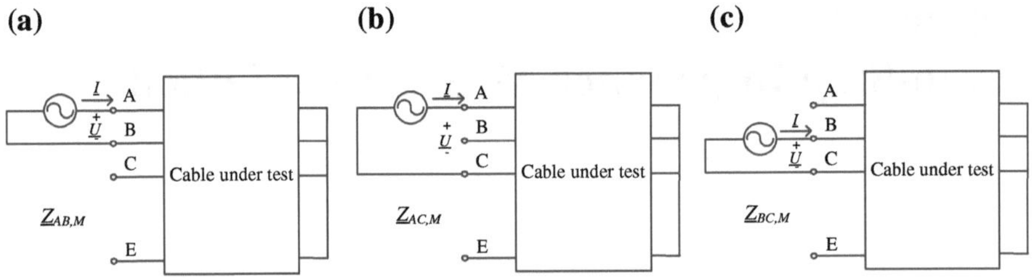

Fig. C.1 Measuring method for determining the positive sequence impedance of a cable system. **a** Impedance between phase A and B, **b** impedance between phase A and C and **c** impedance between phase B and C

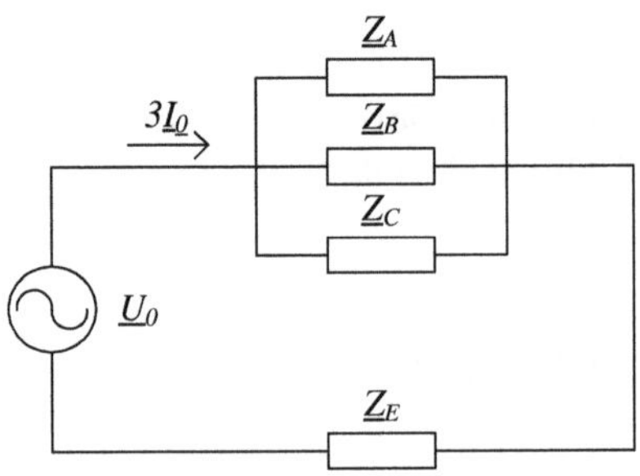

Fig. C.2 Definition of zero sequence impedance system

zero sequence impedance is calculated as:

$$Z_{0-ABC} = \frac{U_{0,M}}{I_{0,M}} = 3Z_{0,M} \qquad [\Omega] \tag{C.2}$$

If the source is not powerful enough to energised all three phases at once, the zero sequence impedance is measured as shown in Fig. C.3a, b and c. The zero-sequence impedance determined using the set up shown in Fig. C.3a is calculated as:

$$Z_{0-A,C} = Z_{A,C} + 3Z_{E-A,C} \qquad [\Omega] \tag{C.3}$$

where $Z_{E-A,C}$ is the return impedance when only phase A is energised. The return impedance $Z_{E-A,C}$ is calculated as:

$$Z_{E-A,C} = Z_{A-E,M} - Z_{A,C} \qquad [\Omega] \tag{C.4}$$

The line impedances $Z_{A,C}$, $Z_{B,C}$ and $Z_{C,C}$ are calculated as:

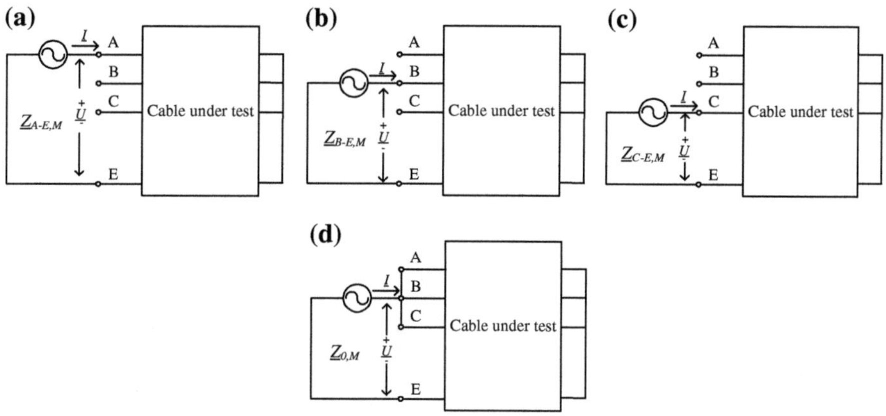

Fig. C.3 Measuring method for determining the zero sequence impedance of a cable system. **a** Impedance between phase A and ground, **b** impedance between phase B and ground and **c** impedance between phase C and ground

$$Z_{A,C} = \frac{Z_{AB,M} - Z_{AC,M} + Z_{BC,M}}{2} \quad [\Omega] \tag{C.5}$$

$$Z_{B,C} = \frac{Z_{AB,M} + Z_{AC,M} - Z_{BC,M}}{2} \quad [\Omega] \tag{C.6}$$

$$Z_{C,C} = \frac{-Z_{AB,M} + Z_{AC,M} + Z_{BC,M}}{2} \quad [\Omega] \tag{C.7}$$

Appendix D
Single Line Diagram of GIS-Station Karstrup

D.1 Single line Diagram of GIS-Station Kastrup

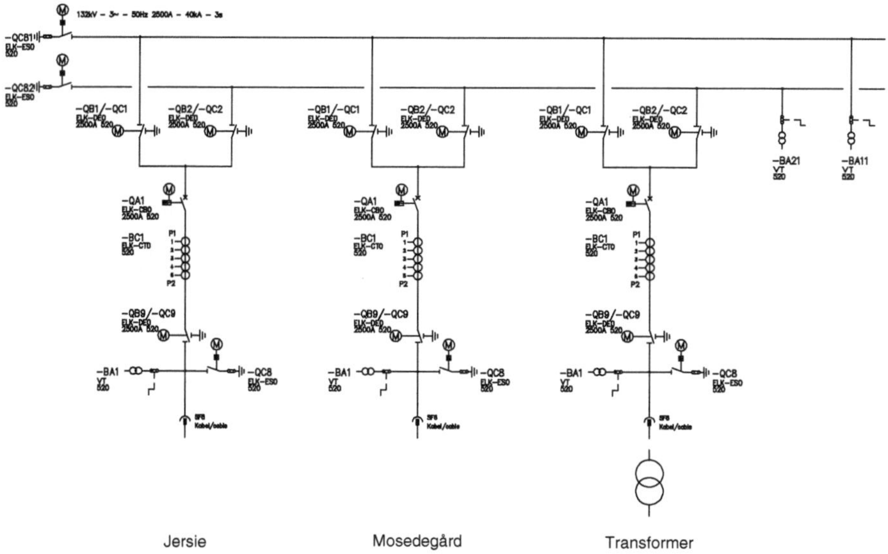

C. F. Jensen, *Online Location of Faults on AC Cables in Underground Transmission Systems*, Springer Theses, DOI: 10.1007/978-3-319-05398-1,

About the Author

Christian Flytkjær Jensen was born in Esbjerg in Denmark, on March 21st, 1980. He studied electrical engineering at Aalborg University in Denmark from where he received both his B.Sc. and M.Sc. degrees. In 2013 he earned the PhD degree also from Aalborg University where his thesis is the foundation for this book.

Christian Flytkjær Jensen is specialised in electrical power systems and high voltage engineering. He is currently employed as an Network Analyst with the Danish transmission system operator Energinet.dk. His main interest are focused on online fault location on undergrounded transmission lines and transient power system modelling.

C. F. Jensen, *Online Location of Faults on AC Cables in Underground Transmission Systems*, Springer Theses, DOI: 10.1007/978-3-319-05398-1,

About the Author

MIX
Papier aus verantwortungsvollen Quellen
Paper from responsible sources
FSC® C105338

If you have any concerns about our products,
you can contact us on
ProductSafety@springernature.com

In case Publisher is established outside the EU,
the EU authorized representative is:
Springer Nature Customer Service Center GmbH
Europaplatz 3, 69115 Heidelberg, Germany

Printed by Libri Plureos GmbH
in Hamburg, Germany